FORTUNE

沛雪立 著

西泠印社出版社

U0652458

图书在版编目（CIP）数据

际遇/ 沛雪立著. -- 杭州: 西泠印社出版社,
2025. 5. -- ISBN 978-7-5508-4837-5

Ⅰ. TQ174.6

中国国家版本馆CIP数据核字第2025S5F533号

际遇

沛雪立　著

选题策划	沛雪立
责任编辑	耿　鑫
责任出版	杨飞凤
责任校对	应俏婷
装帧设计	孙　坤
出版发行	西泠印社出版社

（杭州市西湖文化广场32号5楼　邮政编码　310014）

经　　销	全国新华书店
印　　刷	杭州捷派印务有限公司
开　　本	787mmX1092mm　1/12
字　　数	100千
印　　张	16.5
印　　数	001—800
书　　号	ISBN 978-7-5508-4837-5
版　　次	2025年5月第1版　第1次印刷
定　　价	320.00元

版权所有　翻印必究　印制差错　负责调换

西泠印社出版社发行部联系方式: (0571)87243079

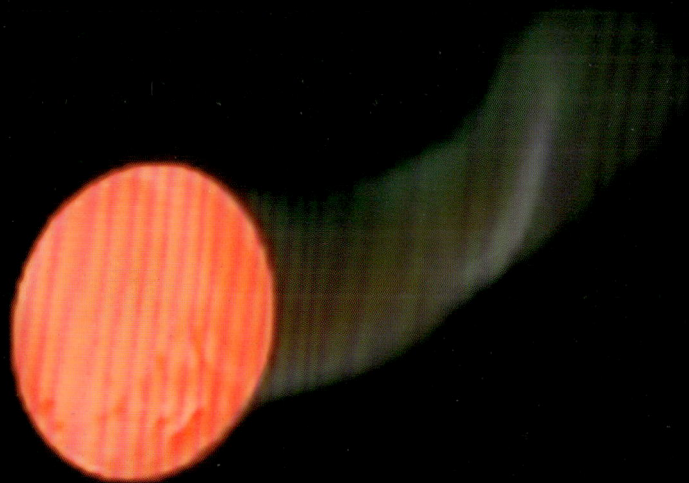

雲在燒沛雪立工作室

雲在燒艺术工作室

青花釉里红的主要原料是氧化钴和氧化铜，烧制过程分为二段：前段是燃除碳素的氧化烧，后段是还原烧。氧化阶段除碳务尽，只有在窑内物质得到充分氧化后，才好干干净净地接受后期的还原。如果氧化不彻底，碳素得不到尽除，会导致还原出现诸多瑕疵。个中的科学原理无须深究，倒是这个氧化阶段的属性颇有意味，两个阶段的关系也似有启示——艺术能干什么？

有如感性与理性的关系。哪些事情是艺术要做且能做的，哪些事情不属于艺术范畴，"僭越"会适得其反。

　　有如氧化与还原的关系。真正能使感觉充分、饱满地进入纯粹表达，正是那个最为珍贵的、弥漫着温存与敏感的、不大靠谱的情绪状态。需要保持的是持续待在这个没有"远见卓识"的状态中，持续专注内心的此时此刻。这个持续的状态好似那个氧化阶段，应该是纯粹的、干净的、彻底的、完全的。

　　好在形式法则或守正、或叛逆、或离经，在当下都能被包容。

　　幸运的是际遇。

沛雪立

2025.1.28

The main raw materials of blue and white with underglaze red are cobalt oxide and copper oxide. The firing process is divided into two stages: the first stage is the oxidation firing to burn off carbon, and the second stage is the reduction firing. In the oxidation stage, all carbon must be completely removed. Only when the substances in the kiln are fully oxidized can they be cleanly and thoroughly reduced in the later stage. If the oxidation is not thorough and the carbon is not completely removed, it will lead to many flaws in the reduction. There is no need to delve into the scientific principles involved. What is more interesting is the nature of the oxidation stage and the relationship between the two stages, which seems to offer some enlightenment — what can art do?

It is like the relationship between sensibility and rationality. What are the things that art should and can do, and what are the things that do not fall within the realm of art? "Overstepping" will be counterproductive.

It is like the relationship between oxidation and reduction. The truly pure, full, and complete expression of feelings comes precisely from that most precious, tender, and sensitive emotional state that is not very reliable. What needs to be maintained is to remain in this state without "foresight" and to focus continuously on the present moment of the inner self. This continuous state is like the oxidation stage, which should be pure, clean, thorough, and complete.

Fortunately, the formal rules, whether they adhere to orthodoxy, rebel, or deviate from the norm, can all be tolerated in the present.

Fortunately, it is an encounter.

Pei Xueli
January 28th, 2025

沛雪立 Pei Xueli

1986年毕业于景德镇陶瓷学院，苏州工艺美术职业技术学院教师，副教授，研讨员级高级工艺美术师，中国美术家协会会员。

Graduated from Jingdezhen Ceramic Institute in 1986, she is a teacher at Suzhou Institute of Arts and Crafts, an associate professor, a senior research fellow and senior level master craftsman of arts and crafts, and a member of the China Artists Association.

青花釉里红自连理以来，演绎出多少千娇百媚。

无所不及的感知，无法遮蔽的真实，认知方式不泥定式的变化，意义的重新发现，表达方式自然无定。

遵循传统工艺美术的装饰法则，折枝、缠枝、适形、对称……在传统装饰意义的结构和骨式法度里，不计一茎一脉、一缘一柄的惟妙惟肖，偏好一笔一墨、一点一线、一晕一染的形态与情绪，倾注真实生活的具体感受，试着把装饰意义转化成表达意义。

Since the inception of blue and white with underglaze red, countless charming variations have emerged.

The all encompassing perception, the unobscured truth, the ever changing cognitive methods that do not adhere to fixed patterns, the rediscovery of meaning, and the natural and fluid forms of expression.

Adhering to the decorative principles of traditional arts and crafts, such as broken branches, entwined branches, adapting to shapes, and symmetry, within the framework and rules of traditional decorative significance, it does not focus on the meticulous imitation of every stem, vein, edge, and handle, but rather prefers the forms and emotions conveyed by every stroke, every ink, every dot, every line, every shade, and every wash. It pours in the concrete feelings of real life, attempting to transform decorative significance into expressive significance.

长：57cm，宽：26cm，高：8cm

宽: 21cm，高: 21cm

16

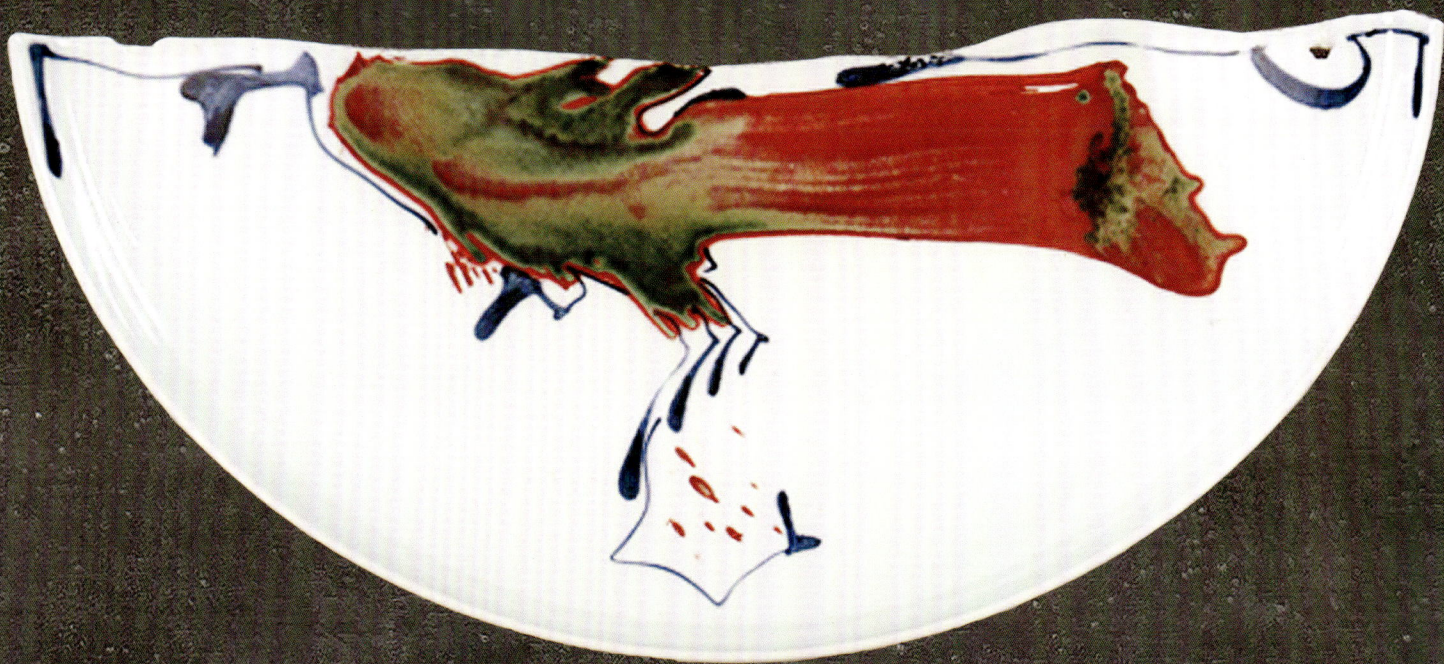

长：57cm，宽：26cm，高：8cm

直径：23cm，高：17cm

直径：8cm，高：21cm

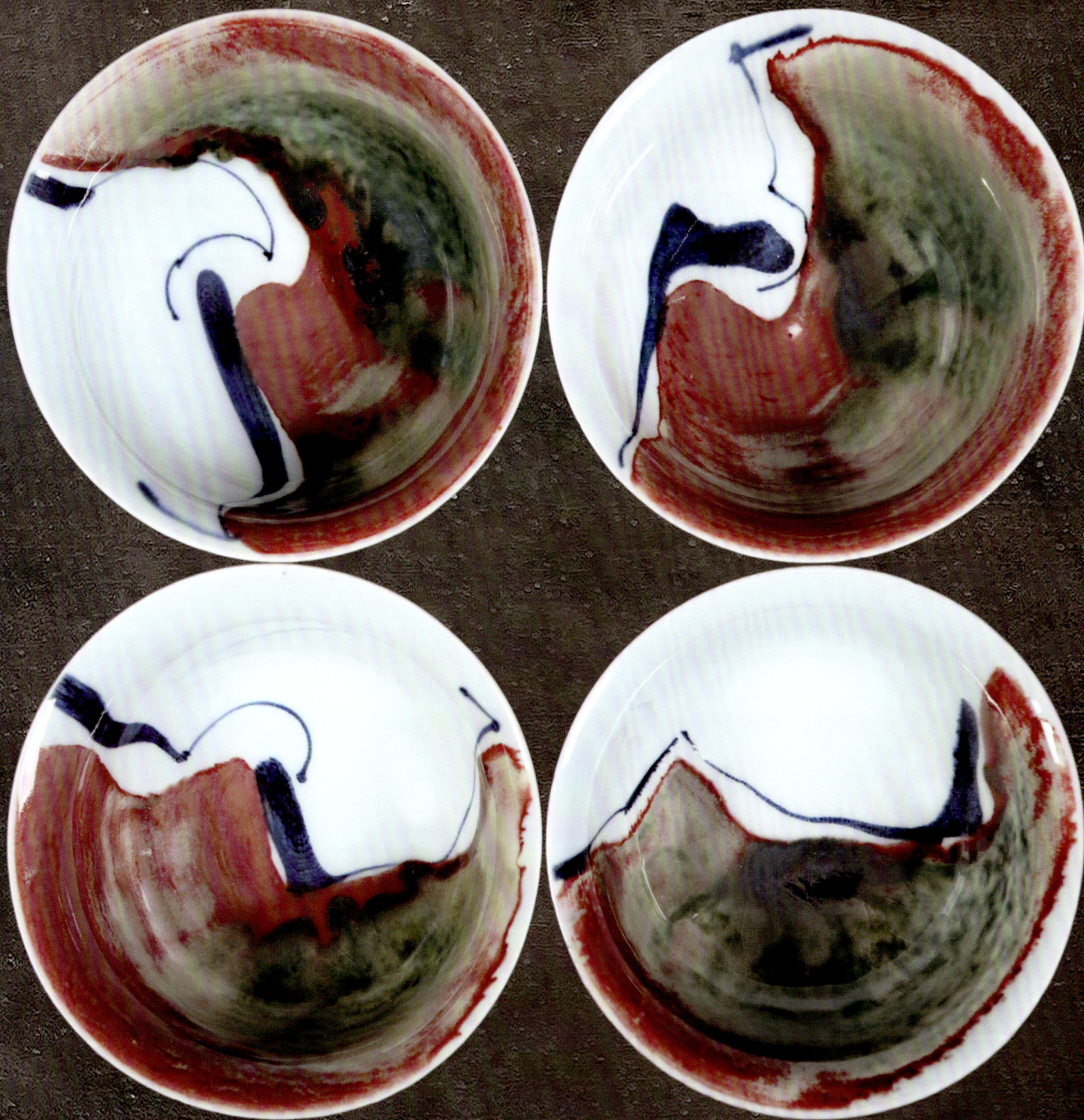

直径: 15cm, 高: 21cm, 数量: 4

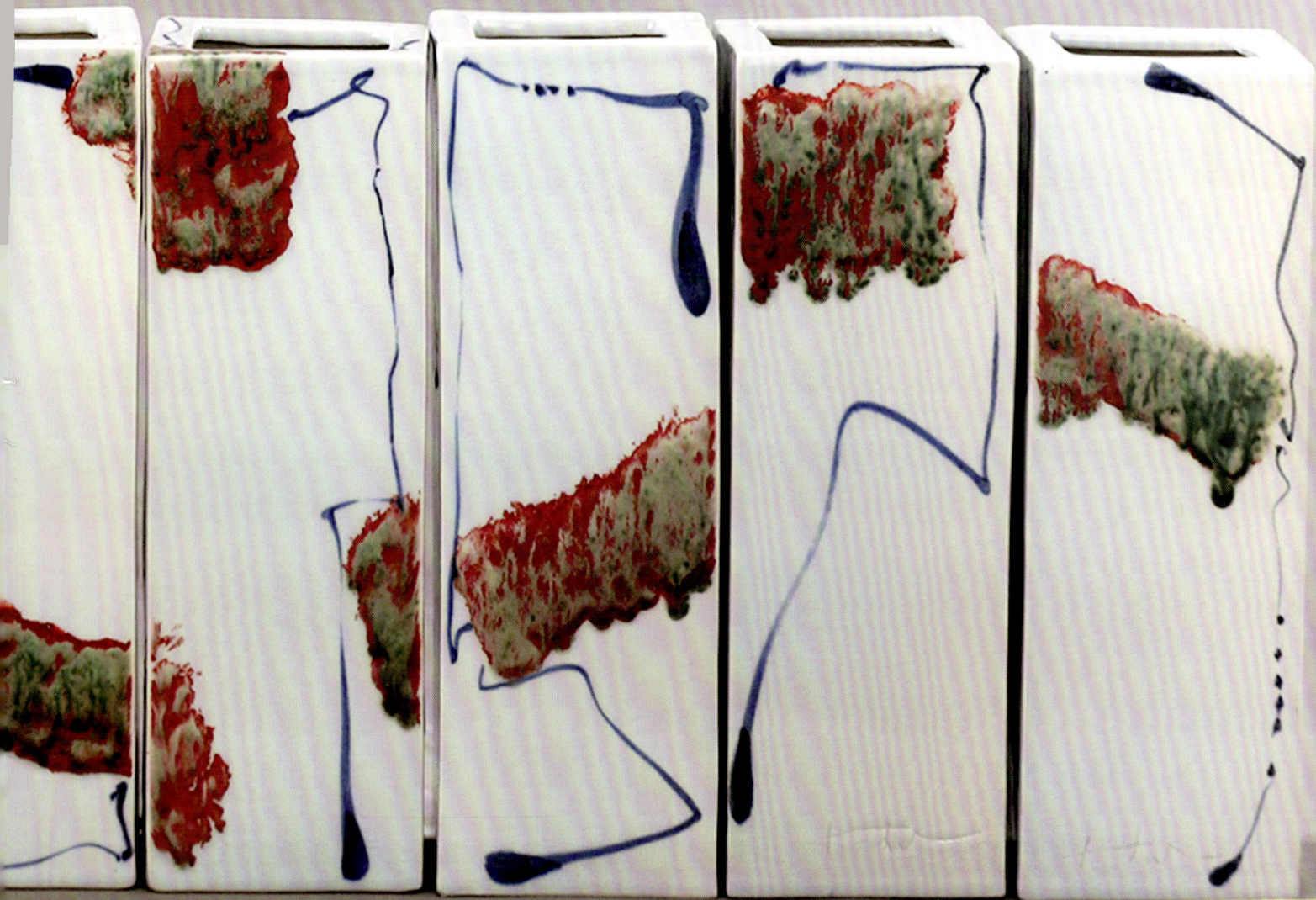

长：11cm，宽：11cm，高：27cm，数量：9

长：70cm，宽：45cm

直径：9cm，高：9cm

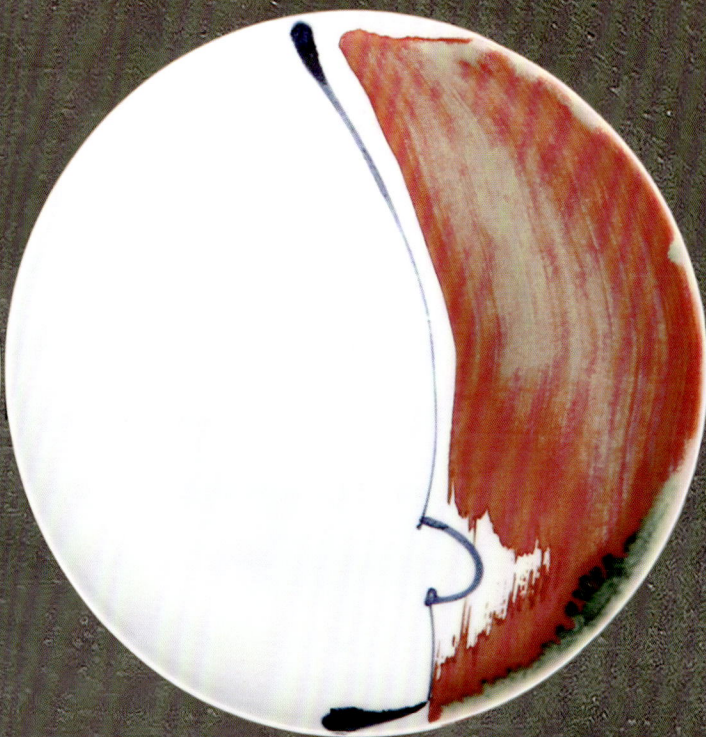

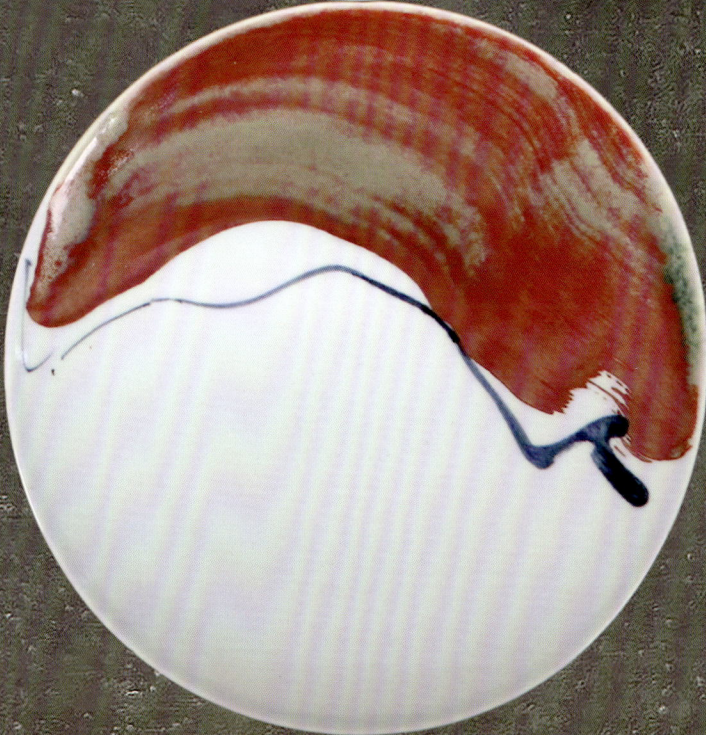

直径：30cm，高：3cm，数量：4

直径：51cm，高：8cm

长：80cm，宽：43cm

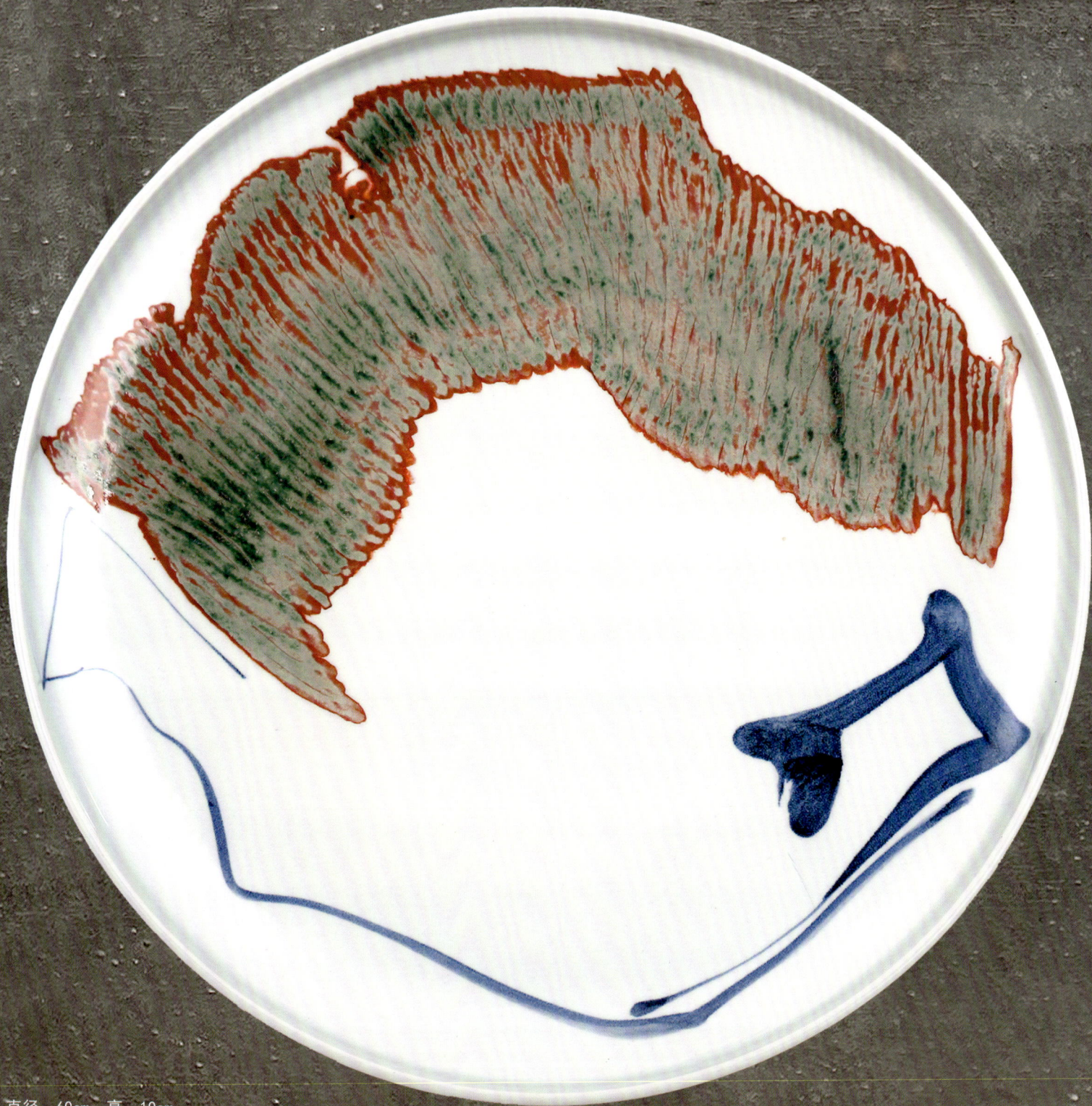

直径：60cm，高：10cm

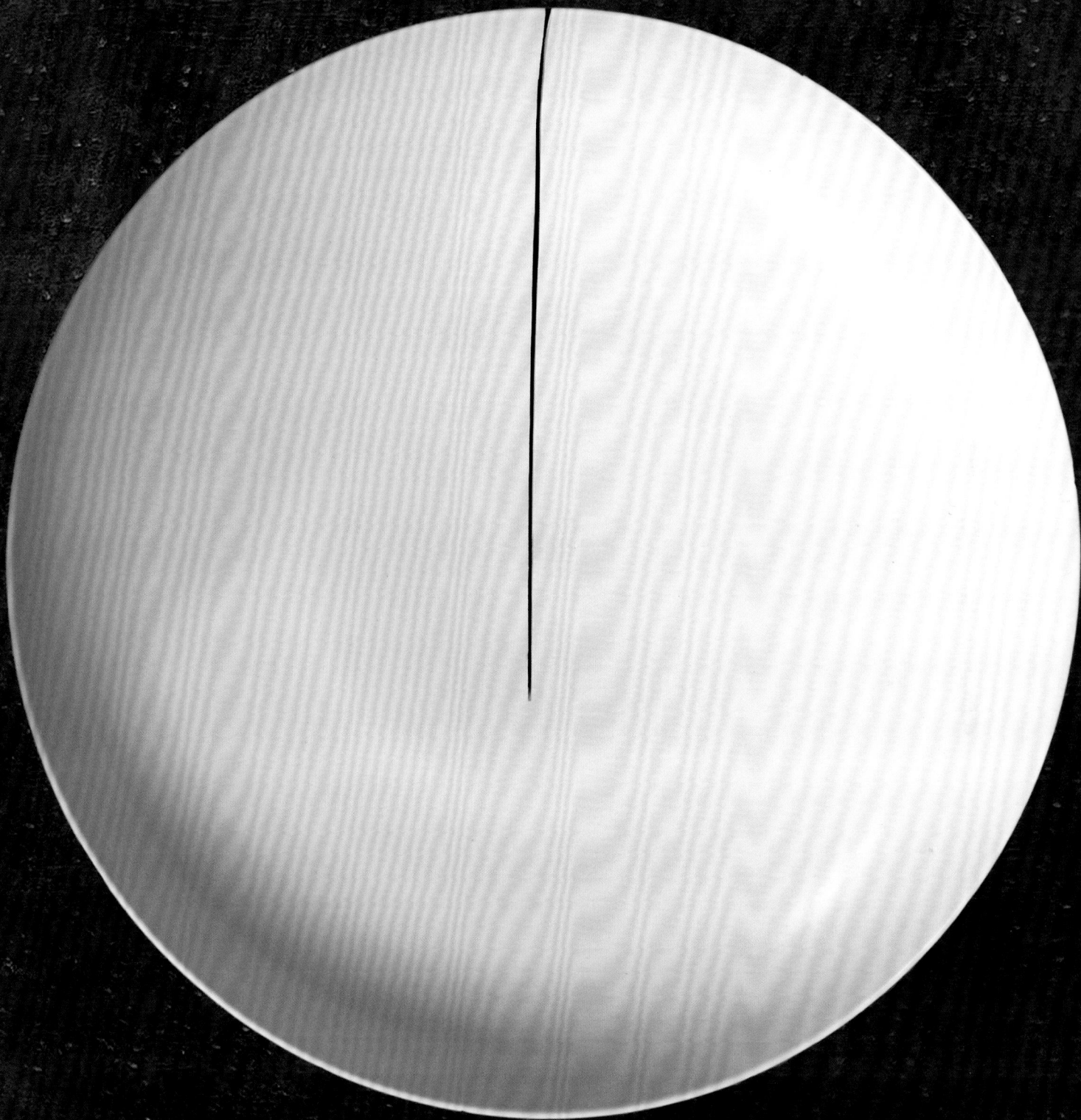

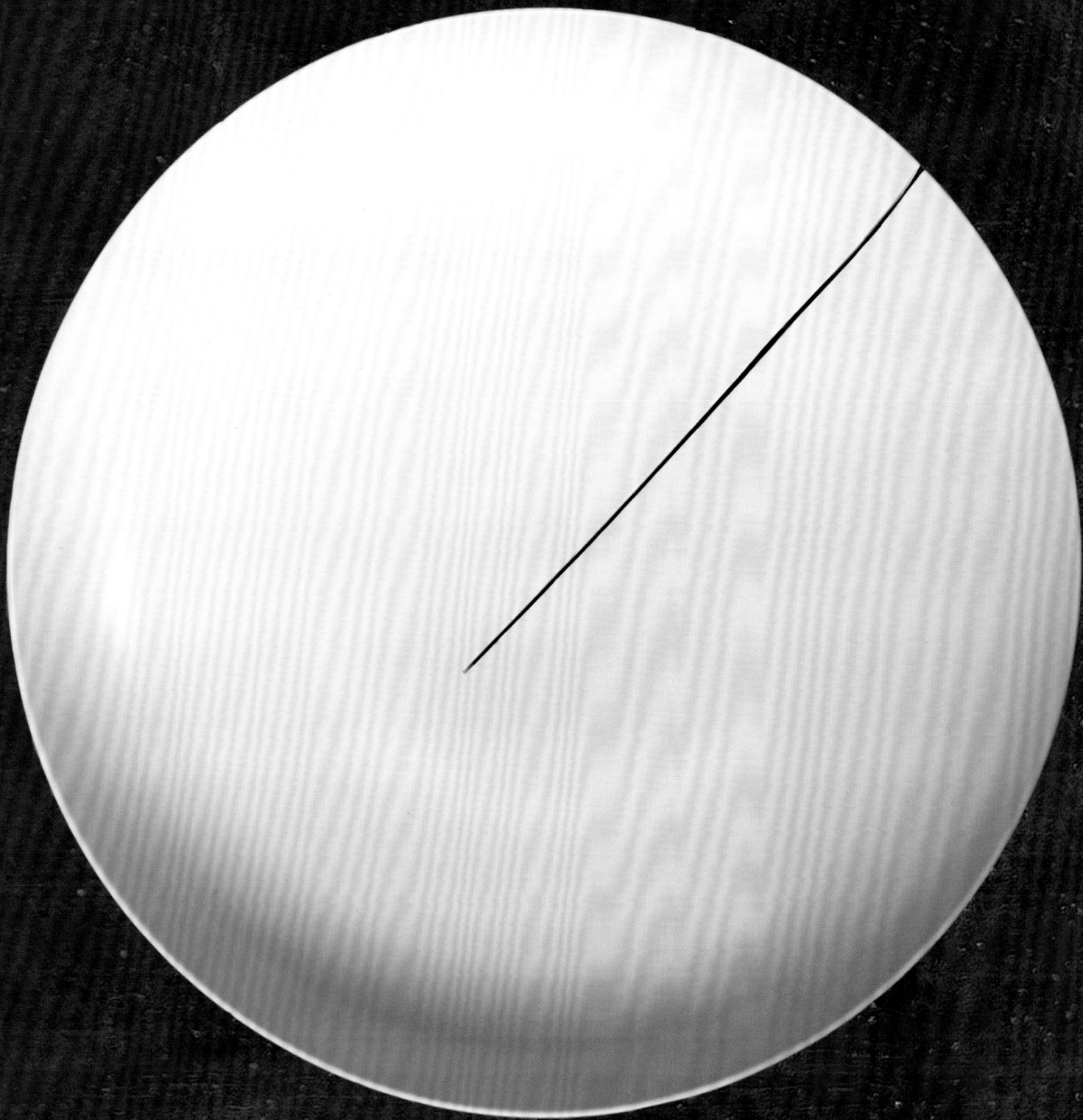

直径：65cm，高：12cm，数量：3

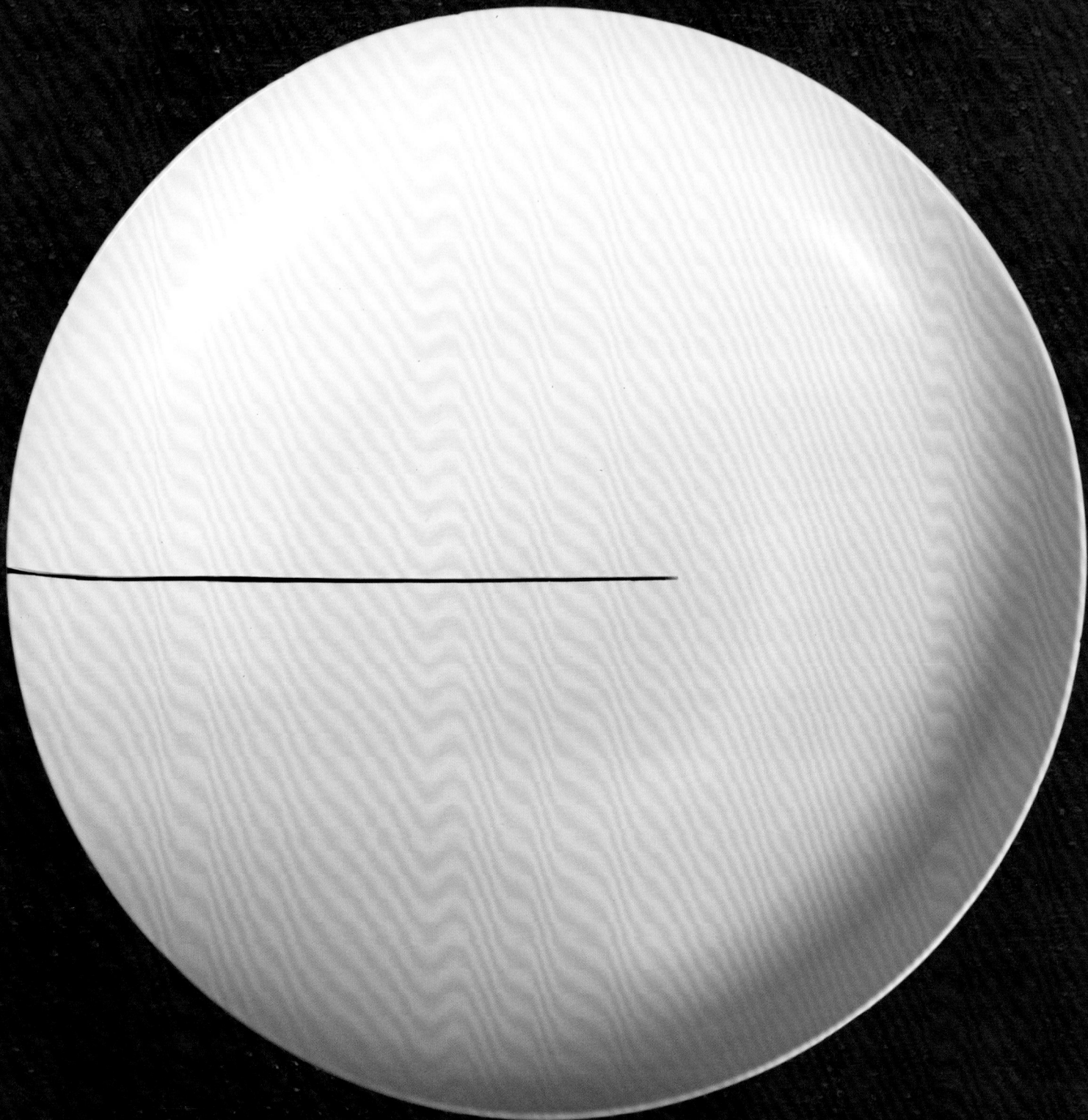

直径：60cm，高：10cm

长：70cm，宽：60cm

长：25cm，宽：25cm，高：77cm

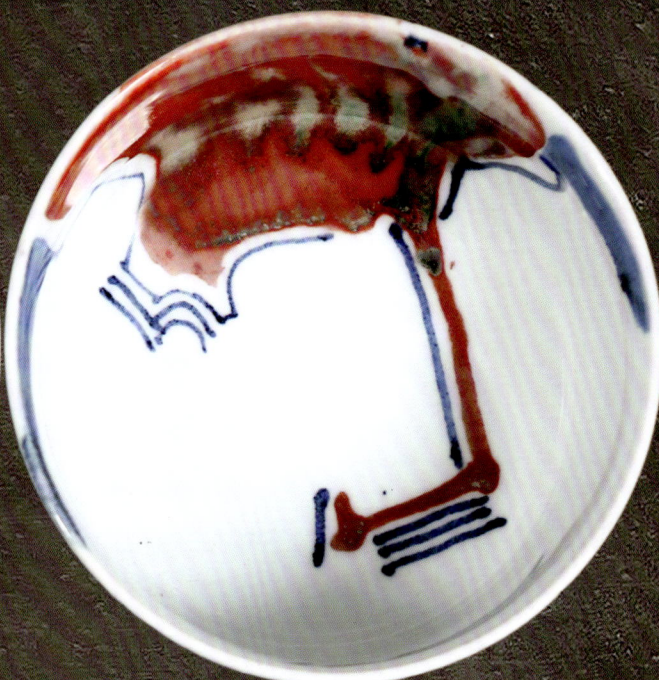

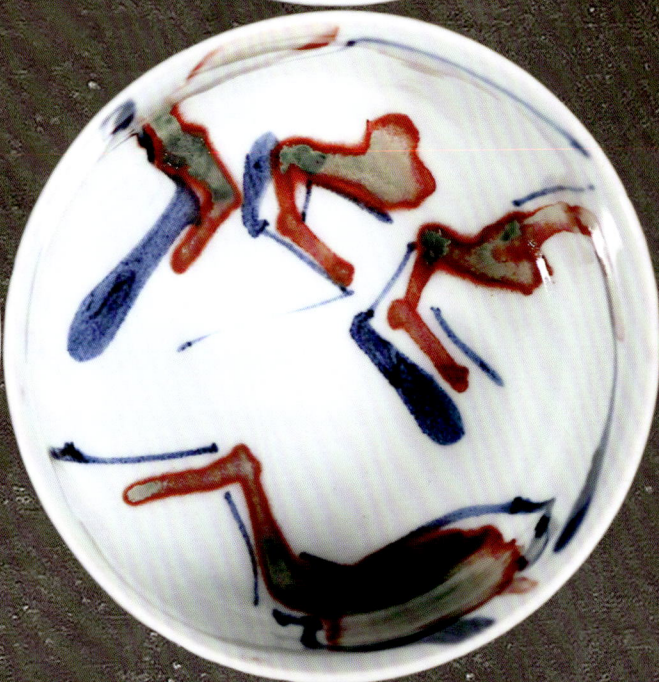

直径：12cm，高：5cm，数量：4

长：80cm，宽：43cm

直径：28cm，高：33cm

直径：51cm，高：8cm

直径：8cm，高：12cm

直径：14cm，高：2cm，数量：4

直径：14cm，高：8cm

直径：6cm，高：7cm

直径：12cm，高：8cm，数量：4

直径：8cm，高：17cm，数量：5

牛仕荣，江西省陶瓷研究所资深工艺师，主研青花、釉里红、高温颜色釉。对红色系的高温烧制技术颇有所得。尤其善于在环境、气候等条件发生较大变化的情况下，对温度、气氛作适时微调，总能像泥里捉鳅般不浓不淡地捉住那份焰气，业界称其为"神气氛"。

Niu Shirong, a senior craftsman at Jiangxi Ceramic Research Institute, mainly focuses on the research of blue and white porcelain, underglaze red, and high temperature colored glazes. He has achieved quite a lot in the high temperature firing technology of red colored glazes. In particular, when there are significant changes in environmental and climatic conditions, he is good at making timely fine adjustments to the temperature and atmosphere. He can always capture the right degree of flame atmosphere just right, like catching loaches in the mud, and the industry calls him the "master of atmosphere".

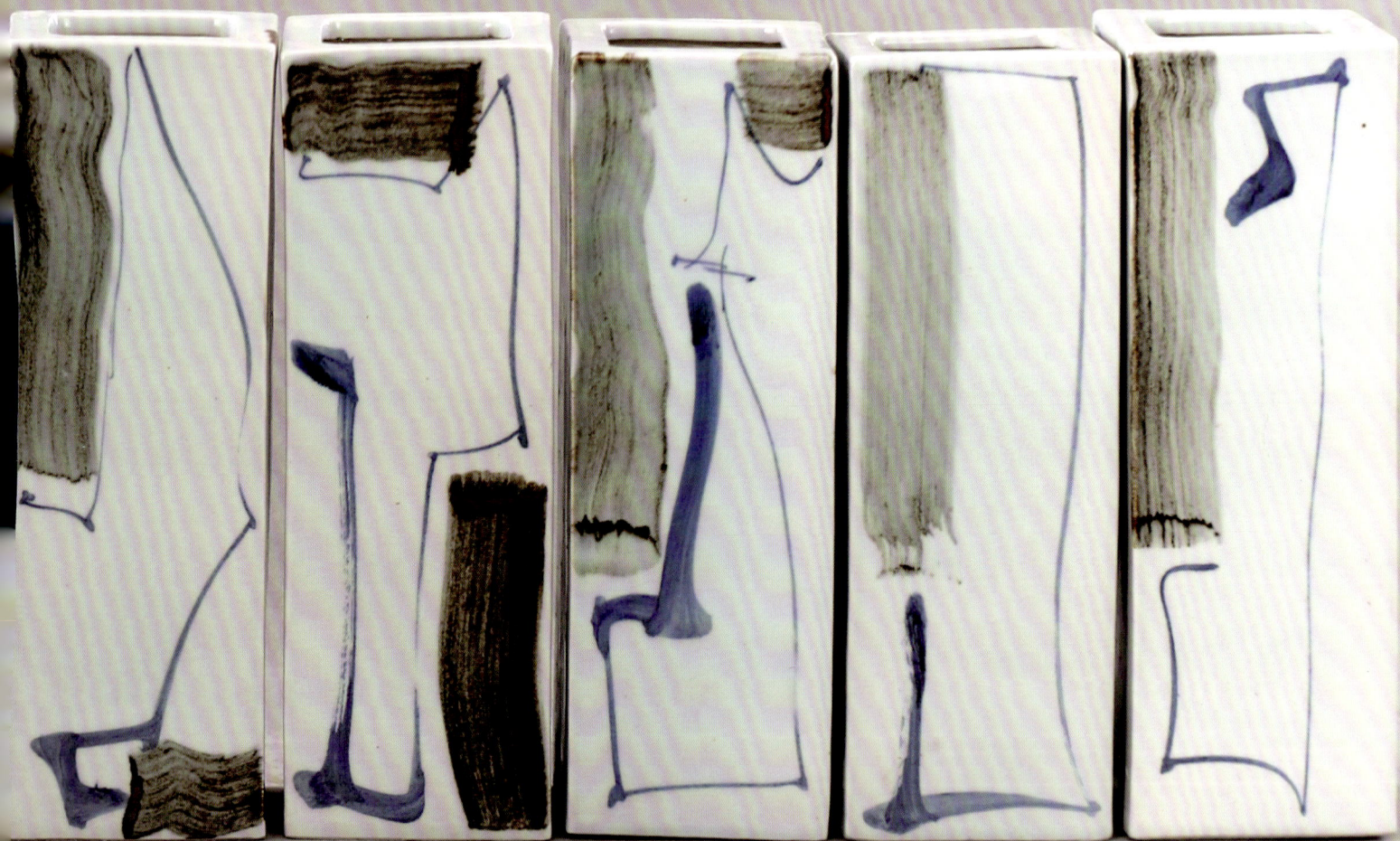

长：11cm，宽：11cm，高：27cm，数量：10

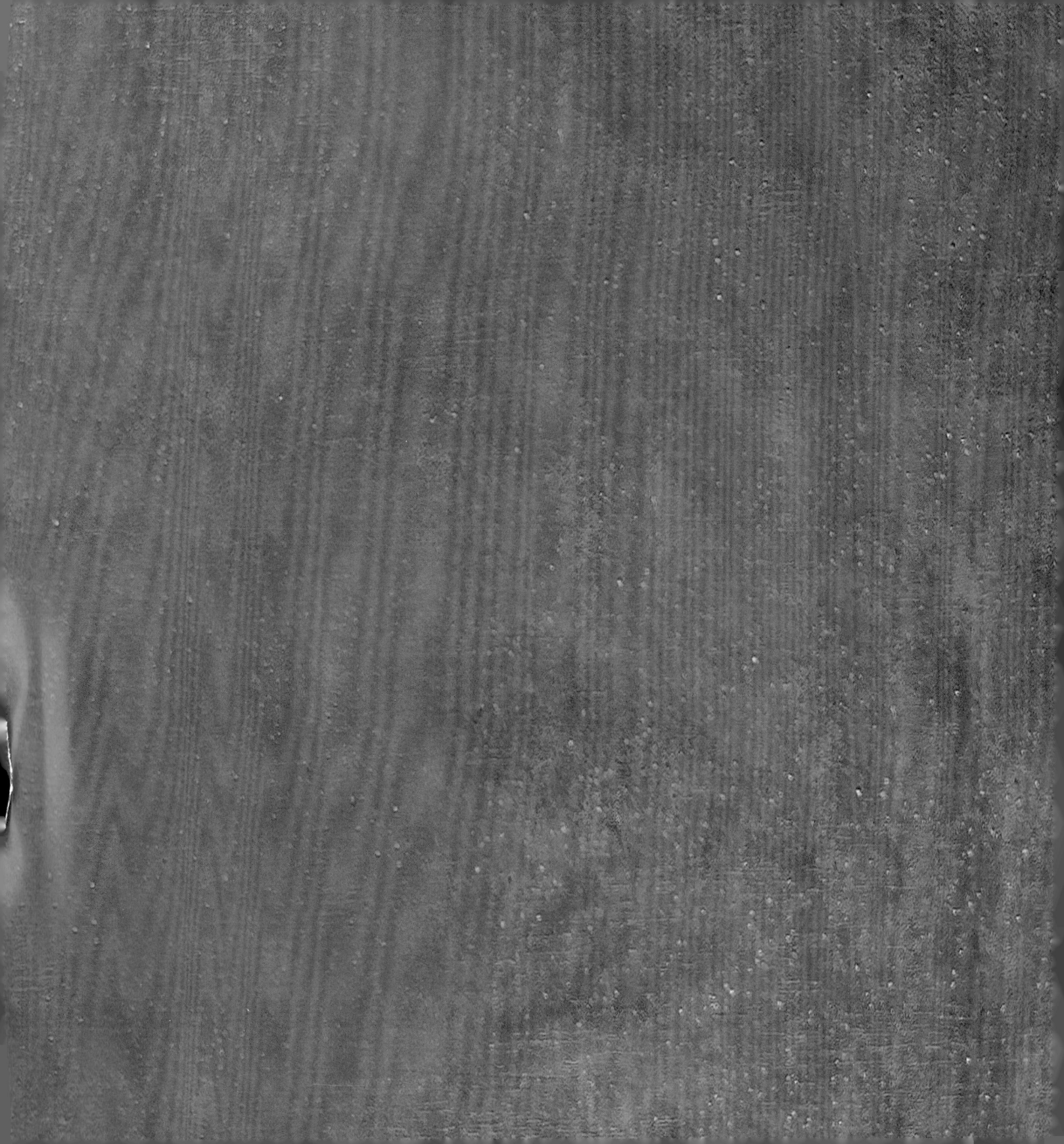

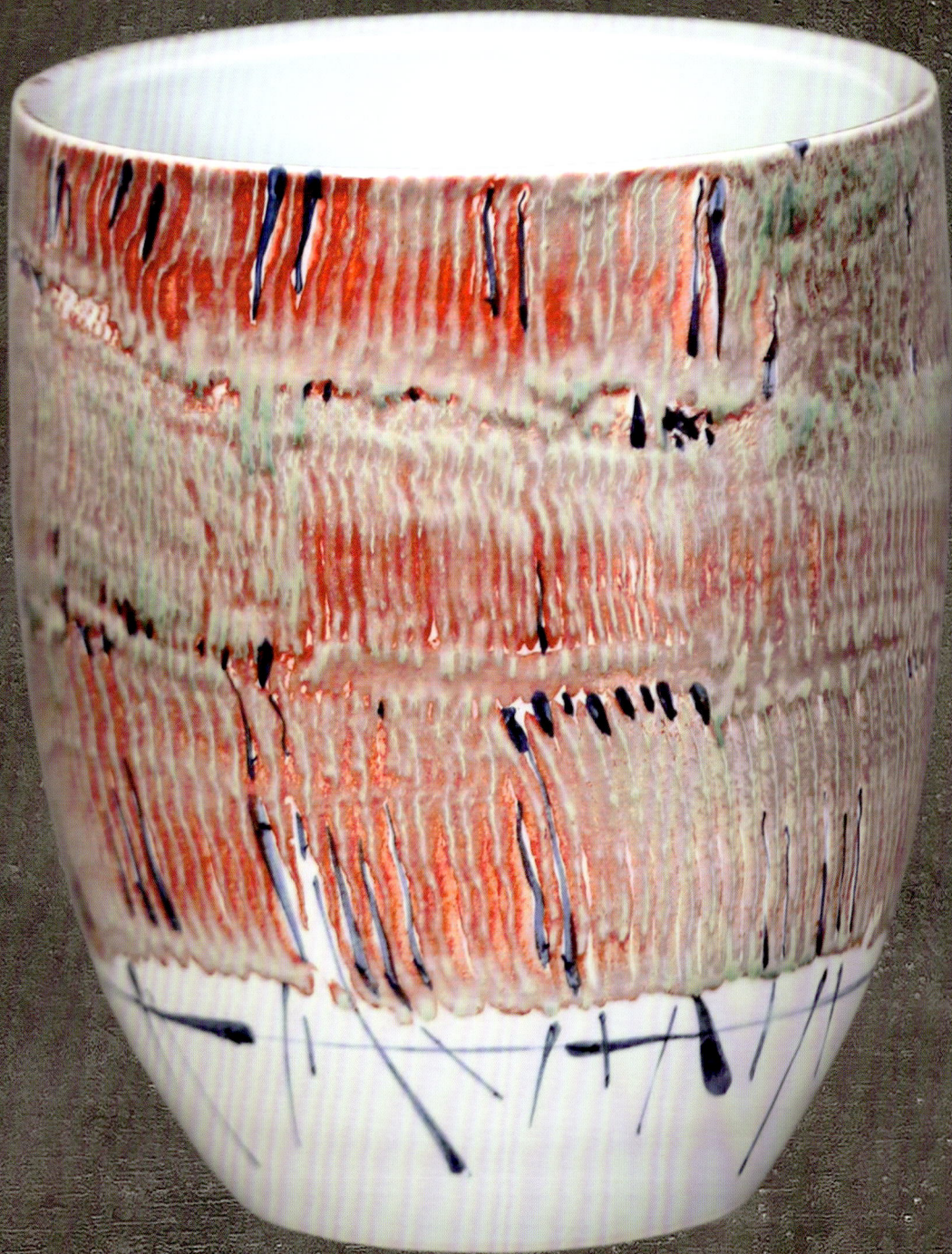

直径：28cm，高：33cm

直径：41cm，高：3cm

直径：14cm，高：3cm，数量：4

直径：6cm，高：6cm，数量：20

直径：31cm，高：3cm

直径：37cm，高：65cm

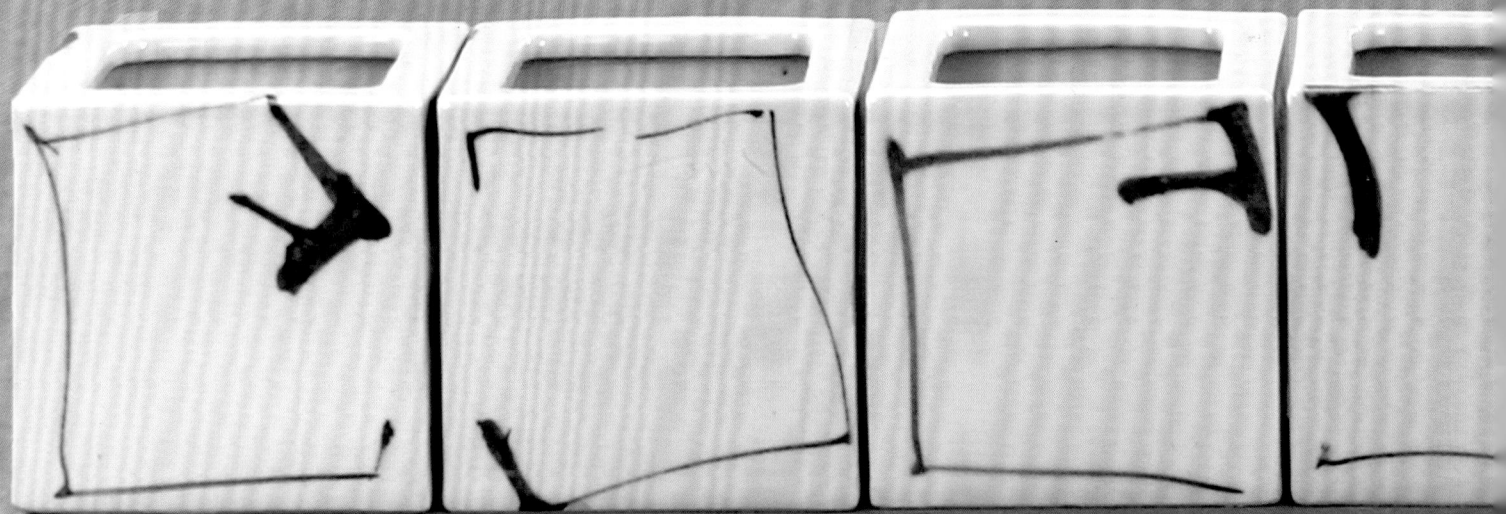

长：12cm，宽：12cm，高：12cm，数量：11

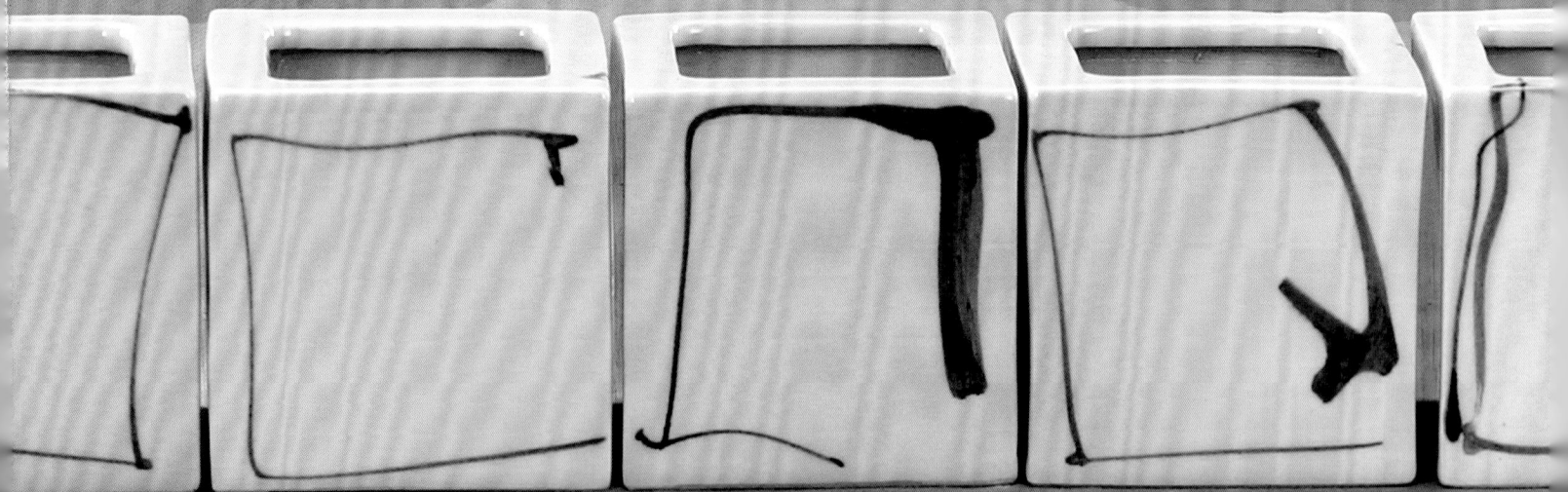

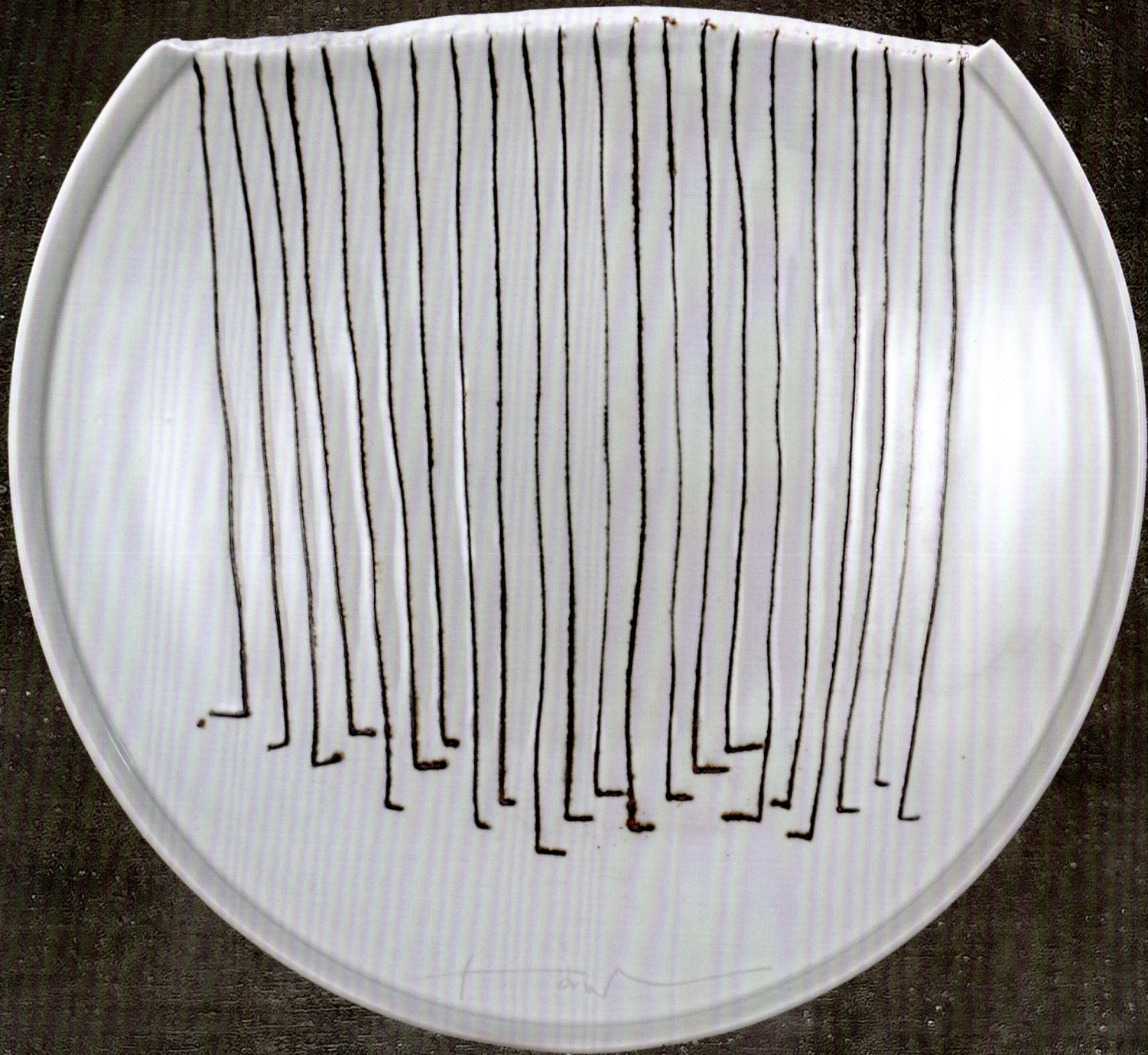

直径：60cm，高：10cm

28cm，高：33cm

直径：48cm，高：8cm

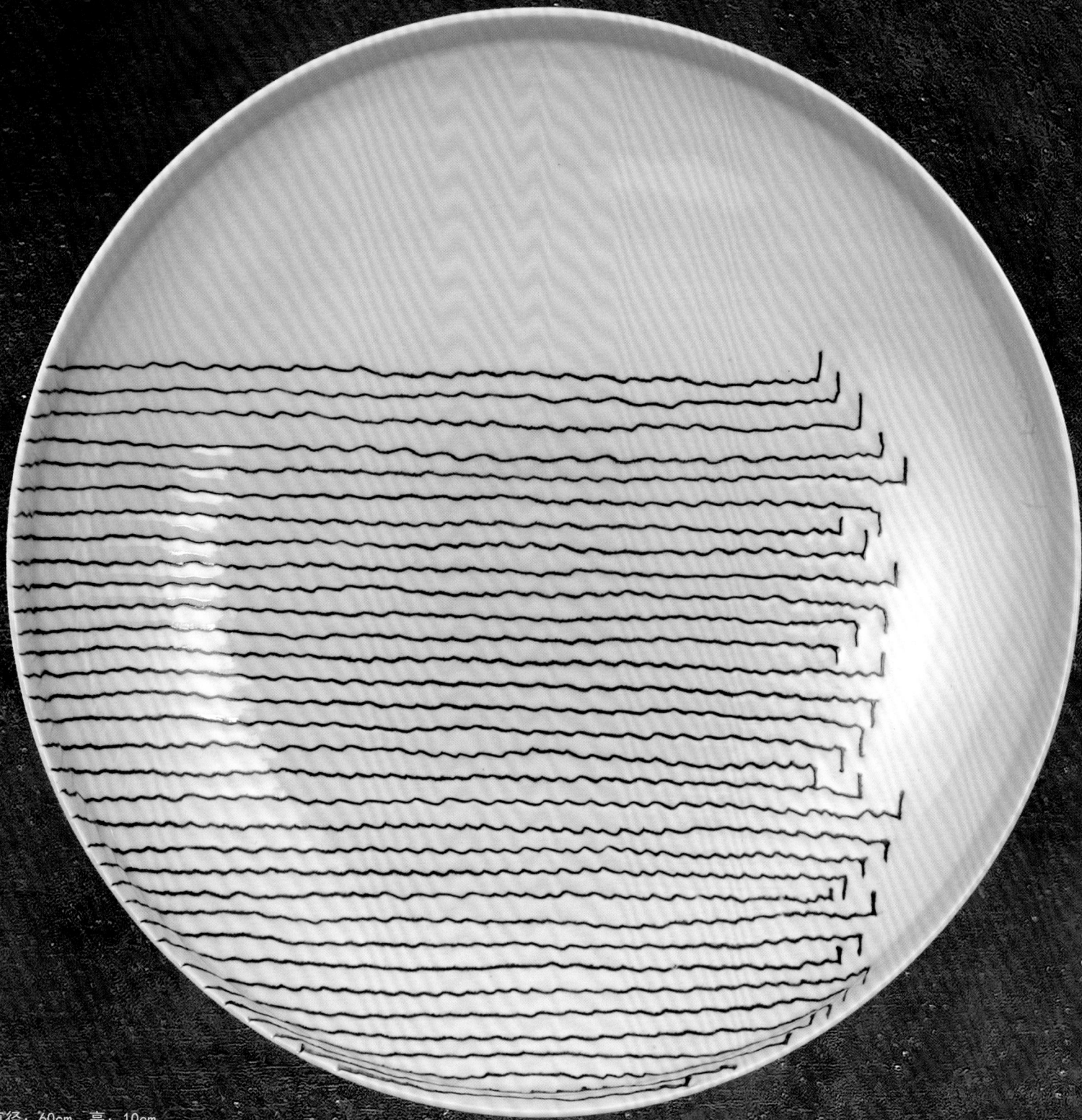

直径：60cm，高：10cm

直径：17cm，高：7cm

直径：28cm，高：33cm

长：8cm，宽：8cm，高：10cm，数量：3

长：8cm，宽：8cm，高：10cm

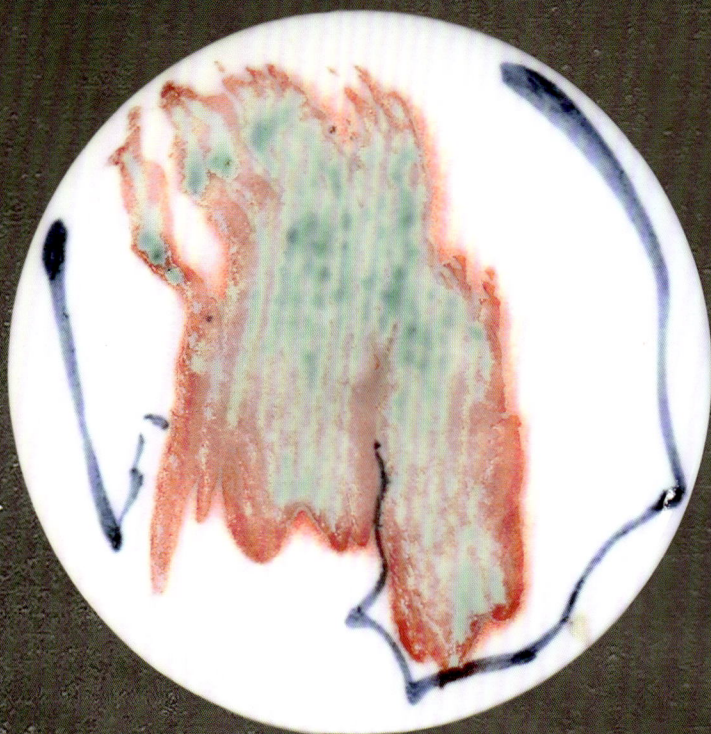

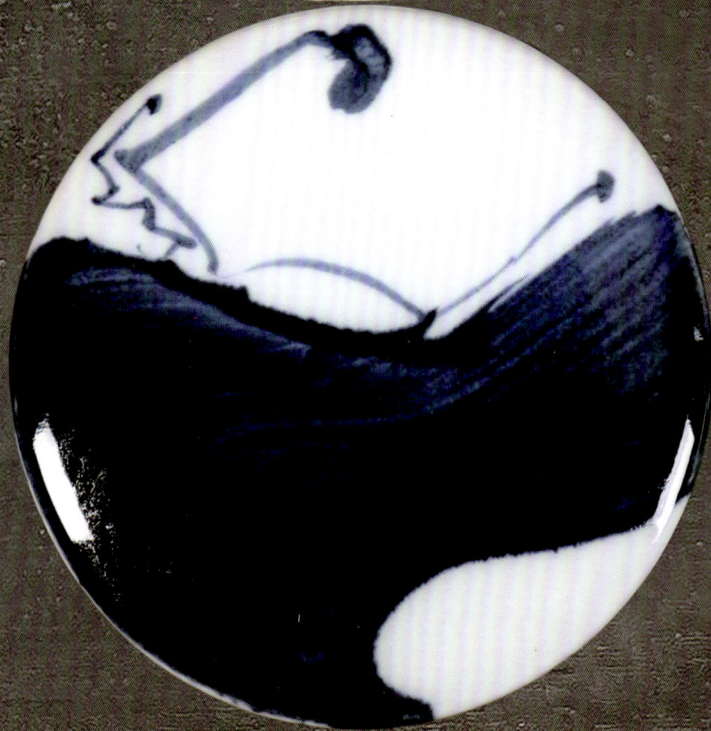

直径：12cm，高：8cm，数量：4

直径：12cm，高：8cm，数量：4

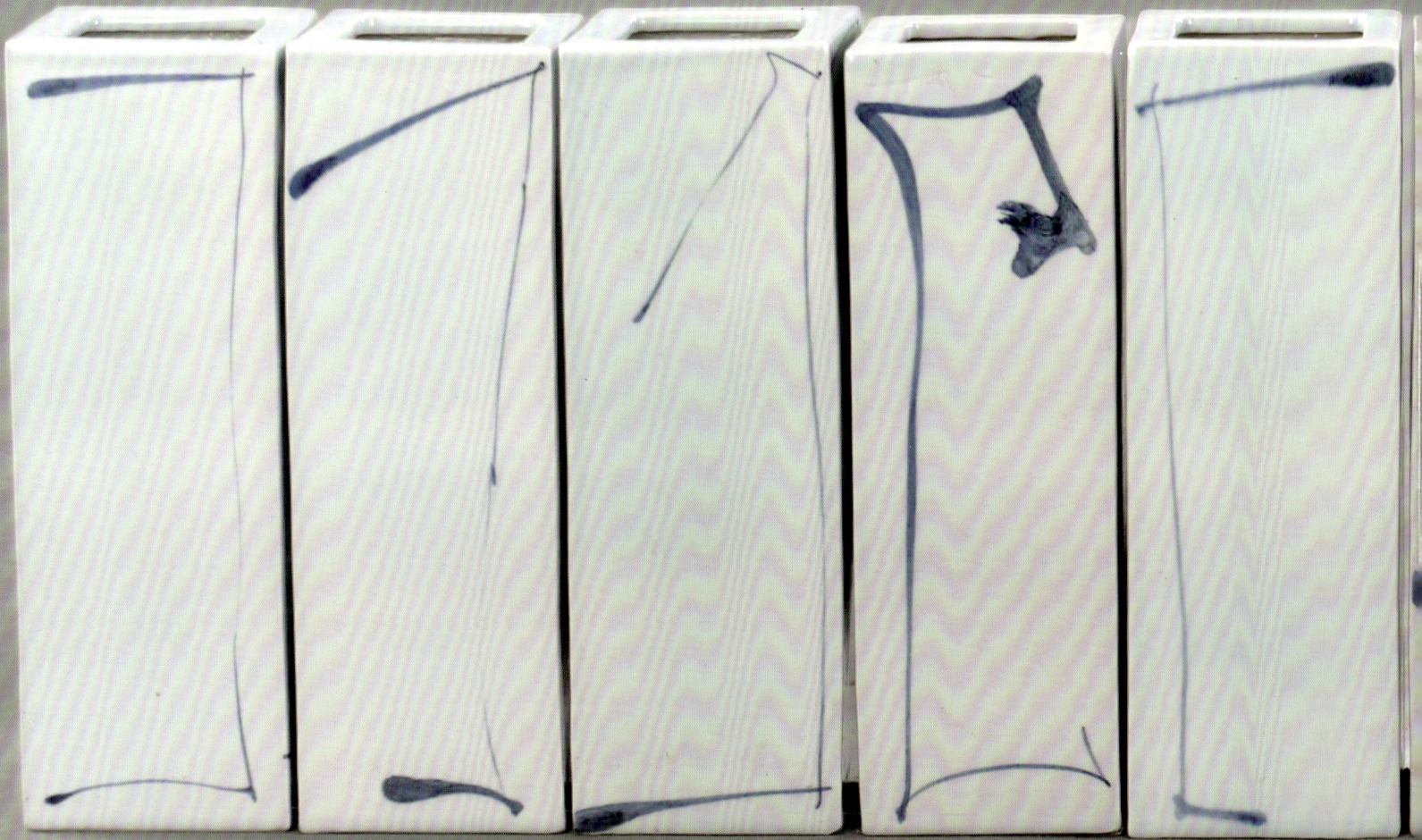

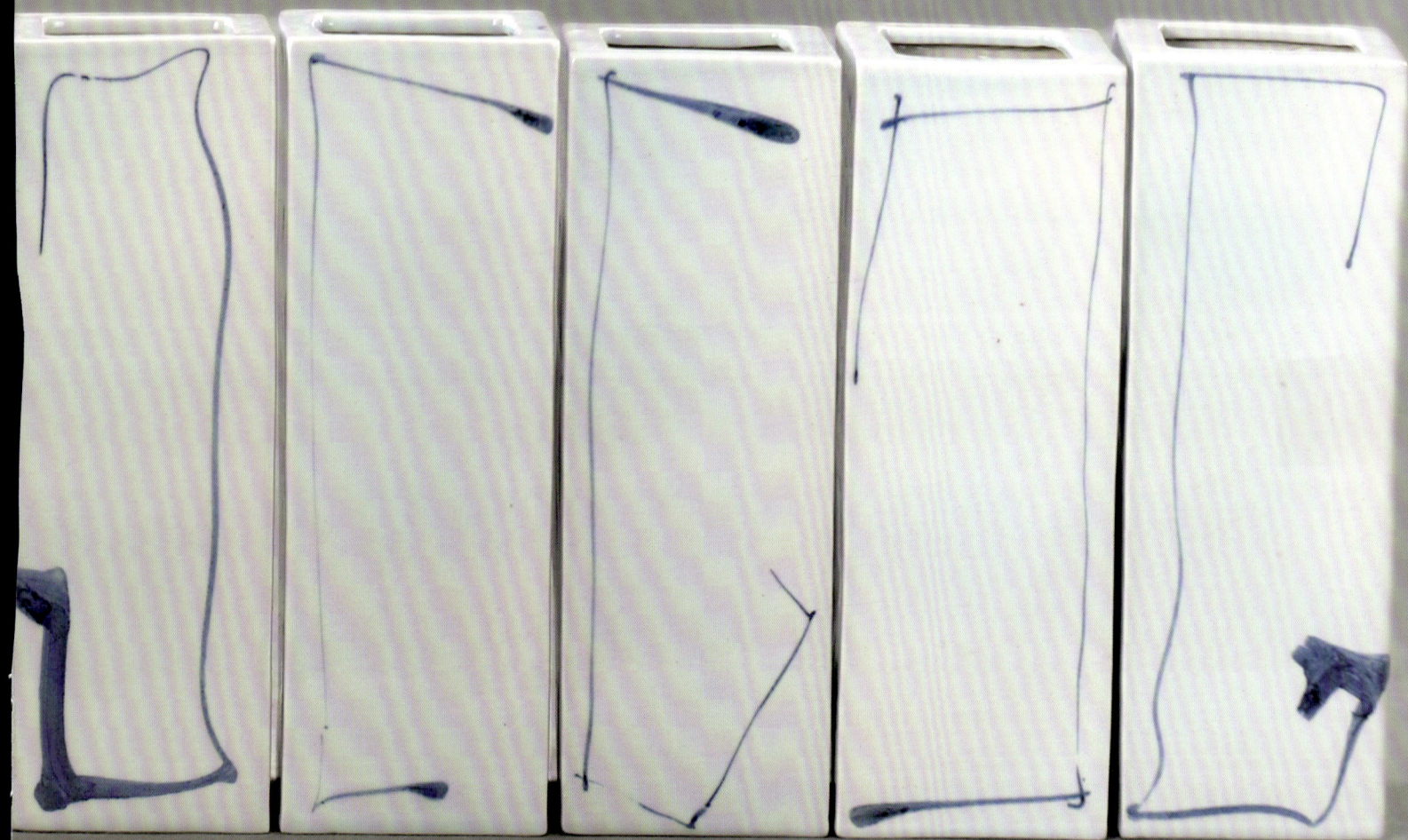

长：11cm，宽：11cm，高：35cm，数量：10

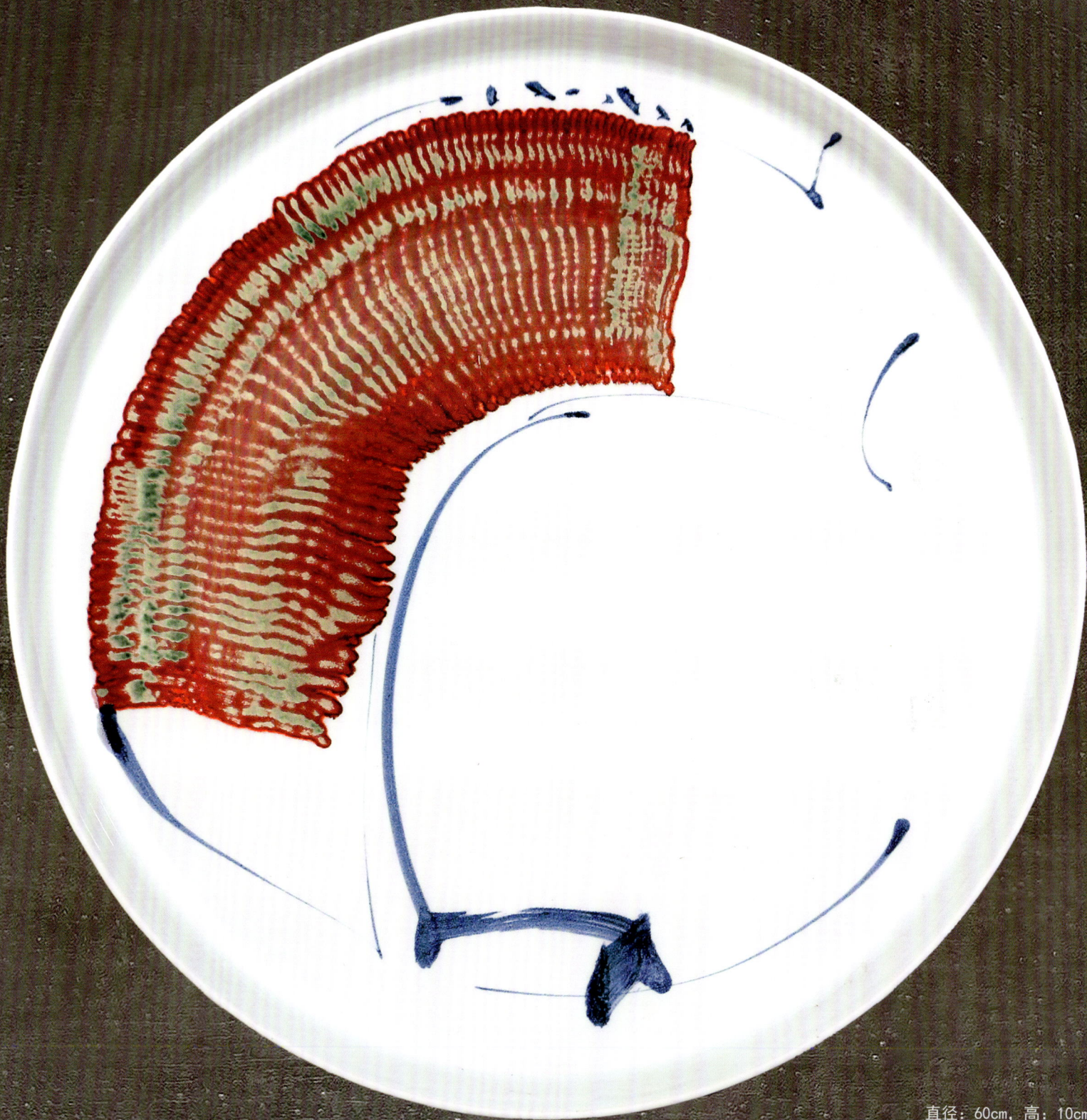

直径：60cm，高：10cm

直径：28cm，高：33cm

直径：48cm，高：8cm

直径：51cm，高：8cm

84

长：180cm，宽：60cm，高：5cm

长：27cm，宽：11cm，高：11cm

直径：28cm，高：33cm

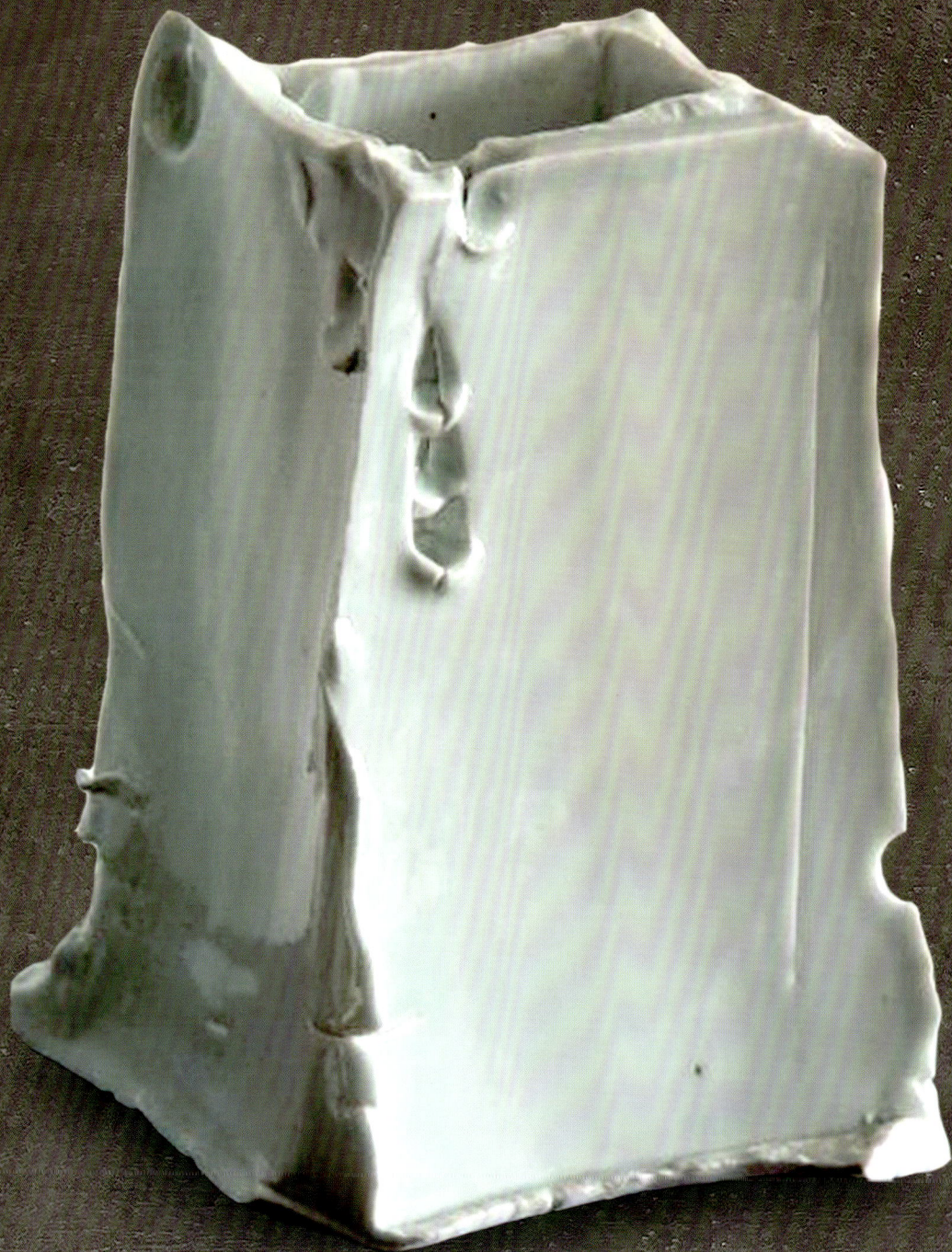

长：18cm，宽：18cm，高：21cm

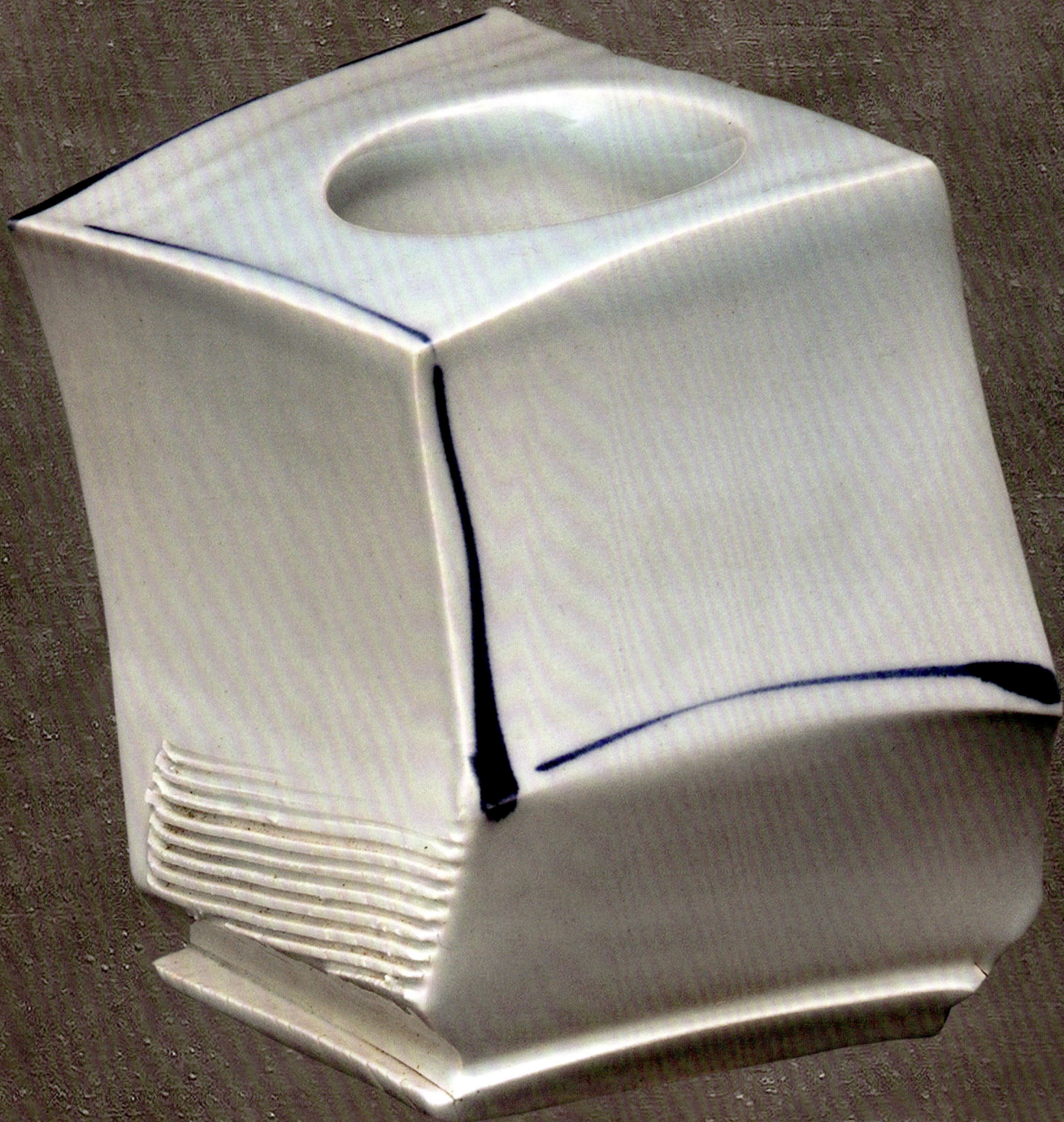

长：13cm，宽：13cm，高：15cm

94

长：15cm，宽：15cm，高：14cm

卫红　Ran Weihong

　　1986年毕业于景德镇陶瓷学院，［□□］工艺美术职业技术学院教师，教授，中国服装设计师协会理事、学术委员会委员，中国流行色协会理事、学术委员会委员。中国工艺美术协会雕塑专业委员会高级会员。

　　Graduated from Jingdezhen Ceramic Institute in 1986. She is a professor at Suzhou Art & Design Technology Institute. She is also a council member and a member of Academic Committee of China Fashion Association, a council member and a member of Academic Committee of China Fashion & Color Association, and a senior member of the Sculpture Professional Committee of China Arts & Crafts Association.

基于性格、心灵、感知、觉悟、沉淀、积累……人的全部，她总试图诠释自然的、社会的、观念的、情绪的、荒诞的、失衡的……凭直觉寻找一切可能性。然而无论怎么辨析明理，都须在一个时间点确定、凝结、物化、呈现。

　　凡事都有价值，忠于内心，改变观看方式，找到心安理得的自洽状态是最高境界。

　　阿尔贝·加缪说："凡是有价值的东西，人都无能为力。"一定意义上这是真理，也是人们不懈创造相对价值的动力。

　　其实，"不可控因素"才是最宝贵的。

Based on one's entire being, including personality, soul, perception, awareness, precipitation, and accumulation, she attempts to interpret the natural, social, ideological, emotional, absurd, and imbalanced aspects, and seeks all possibilities by intuition. However, no matter how clearly one analyzes and understands, it is necessary to determine, condense, materialize, and present at a certain point in time.

Everything has value. Being true to one's heart, changing the way of looking, and finding a state of self consistency with a clear conscience is the highest realm.

Albert Camus said, "Man is powerless in the face of everything that is of value." In a certain sense, this is the truth and also the driving force for people to unremittingly create relative value.

In fact, the "uncontrollable factors" are the most precious.

直径：13cm，高：2cm，数量：36

直径：51cm，高：8cm

102

长：180cm，宽：50cm

长：150cm，宽：60cm

长：110cm，宽：90cm

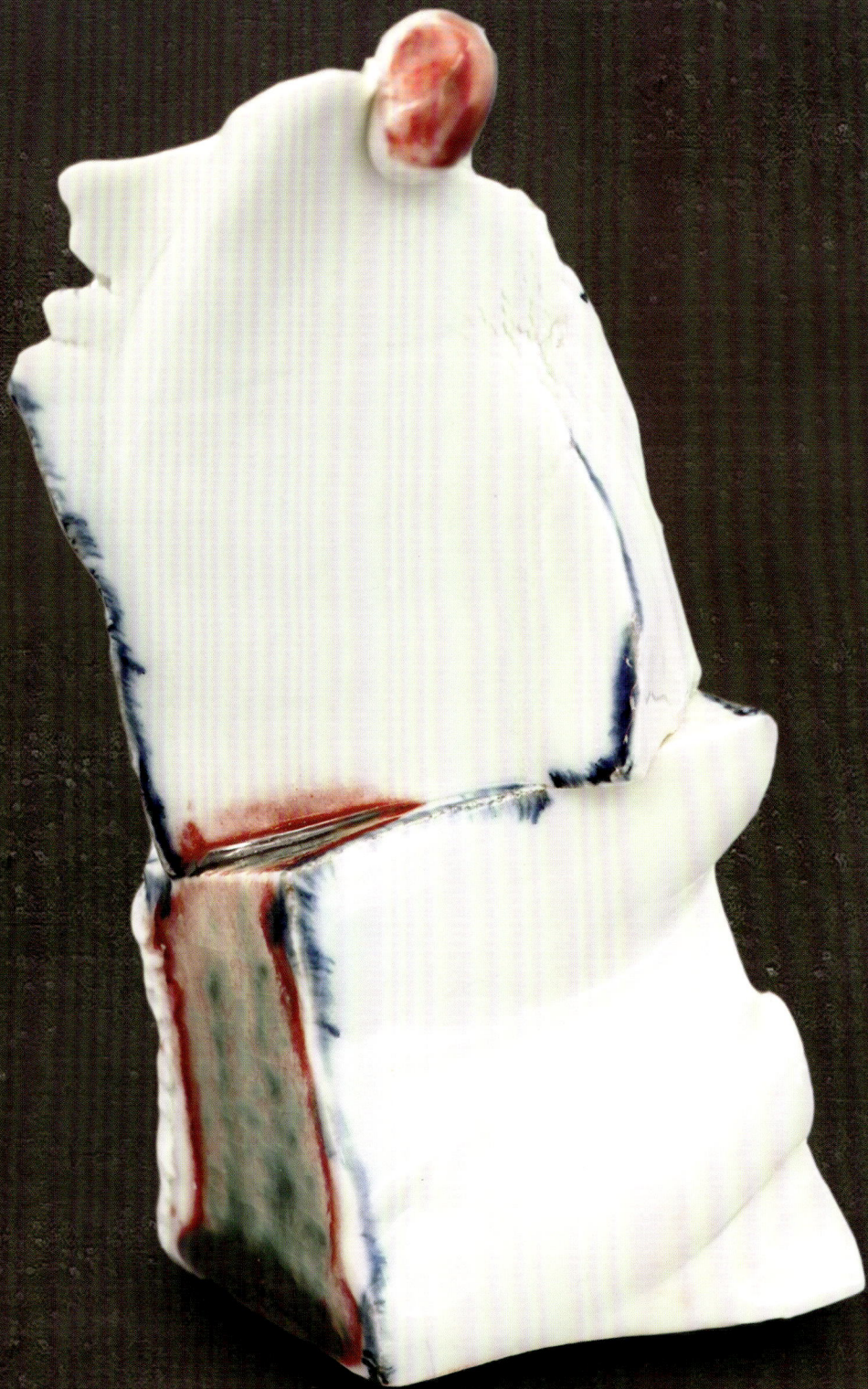

长：10cm，宽：10cm，高：18cm

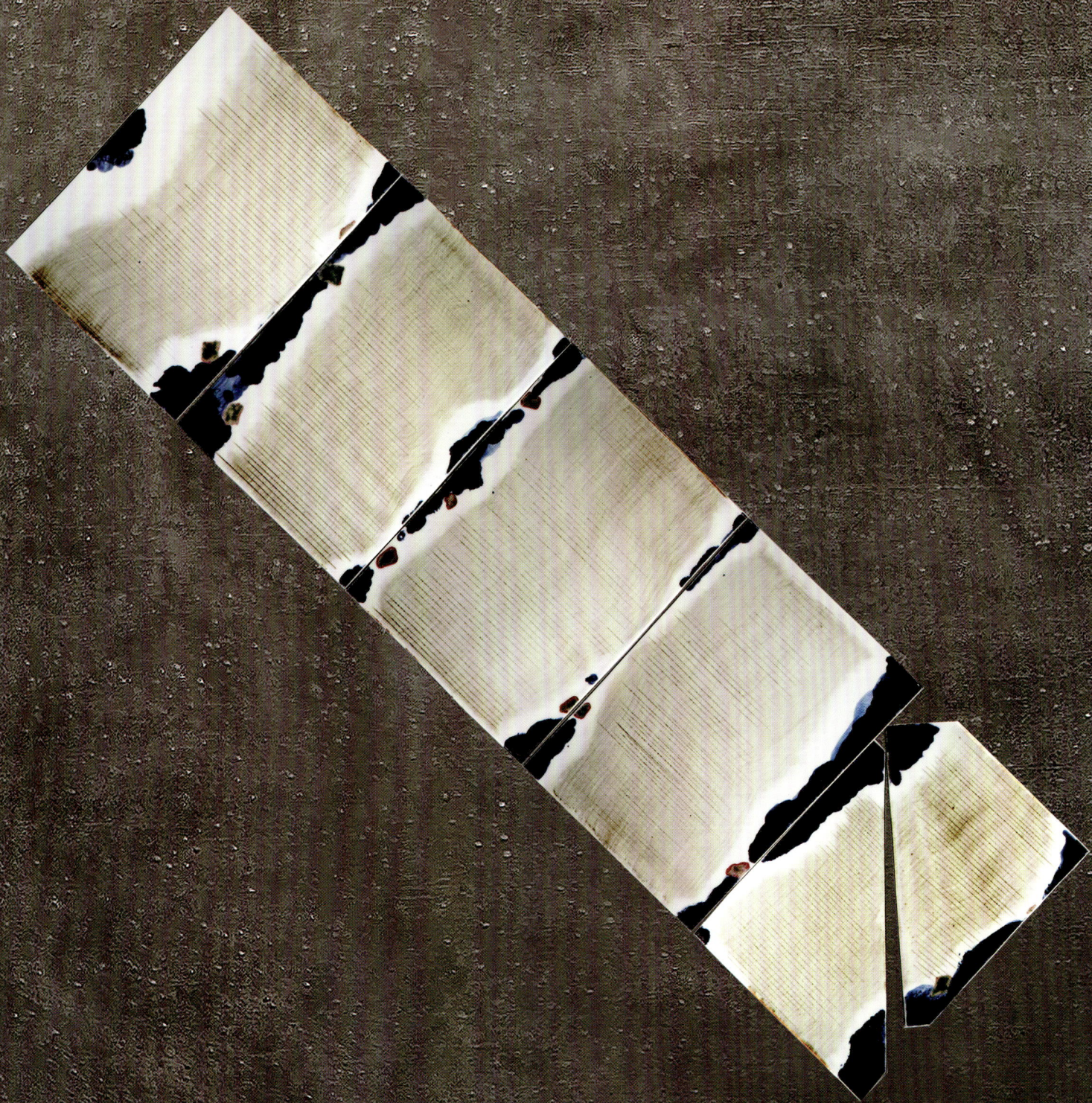

长：150cm，宽：48cm

直径：13cm，高：2cm，数量：80

长：100cm，宽：60cm

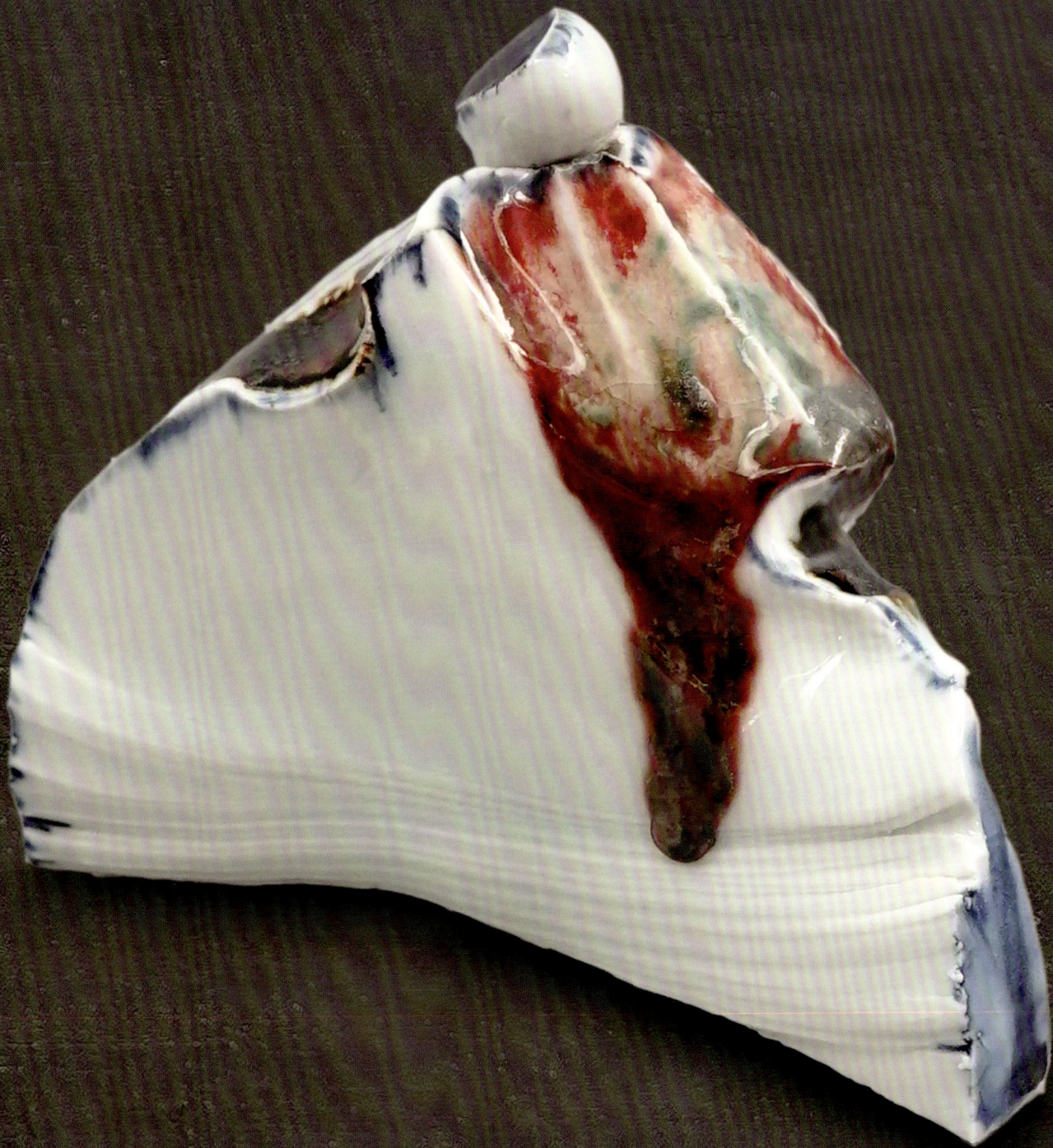

长：16cm，宽：6cm，高：16cm

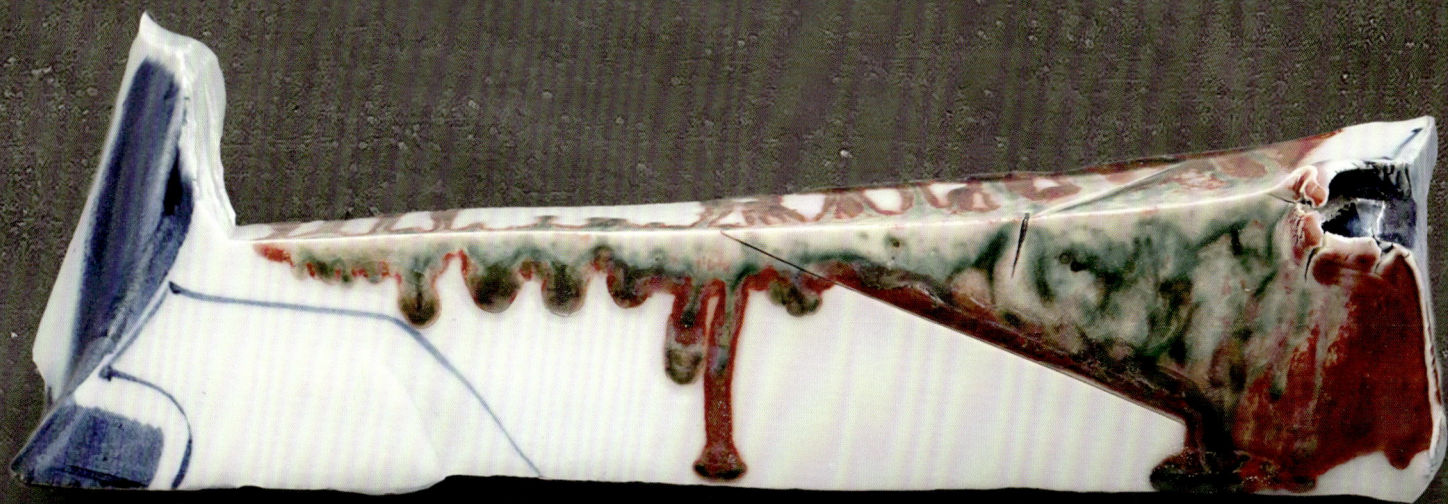

长：30cm，宽：10cm，高：9cm

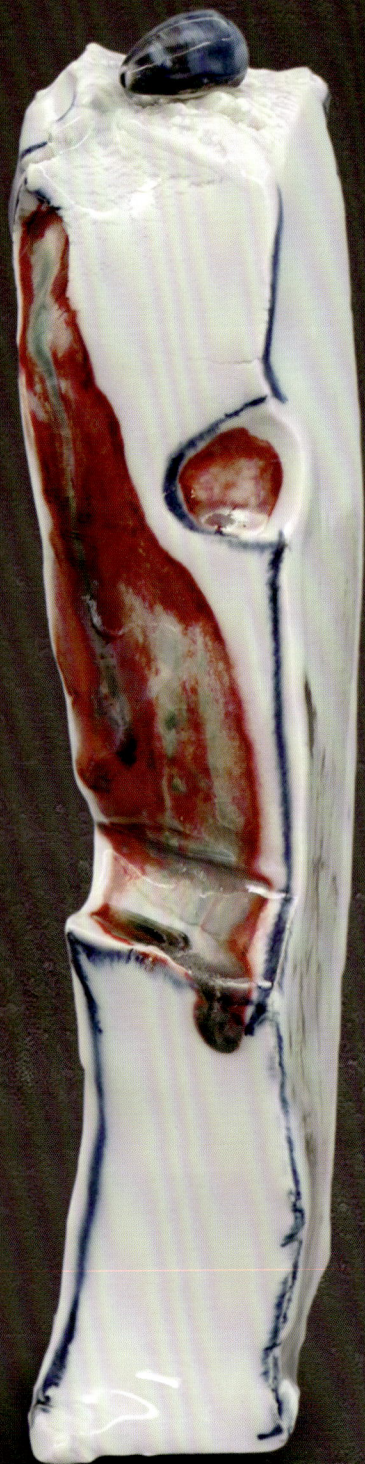

长：10cm，宽：10cm，高：30cm

长：26cm，宽：13cm，高：20cm

长：13cm，宽：13cm，高：21cm

60cm

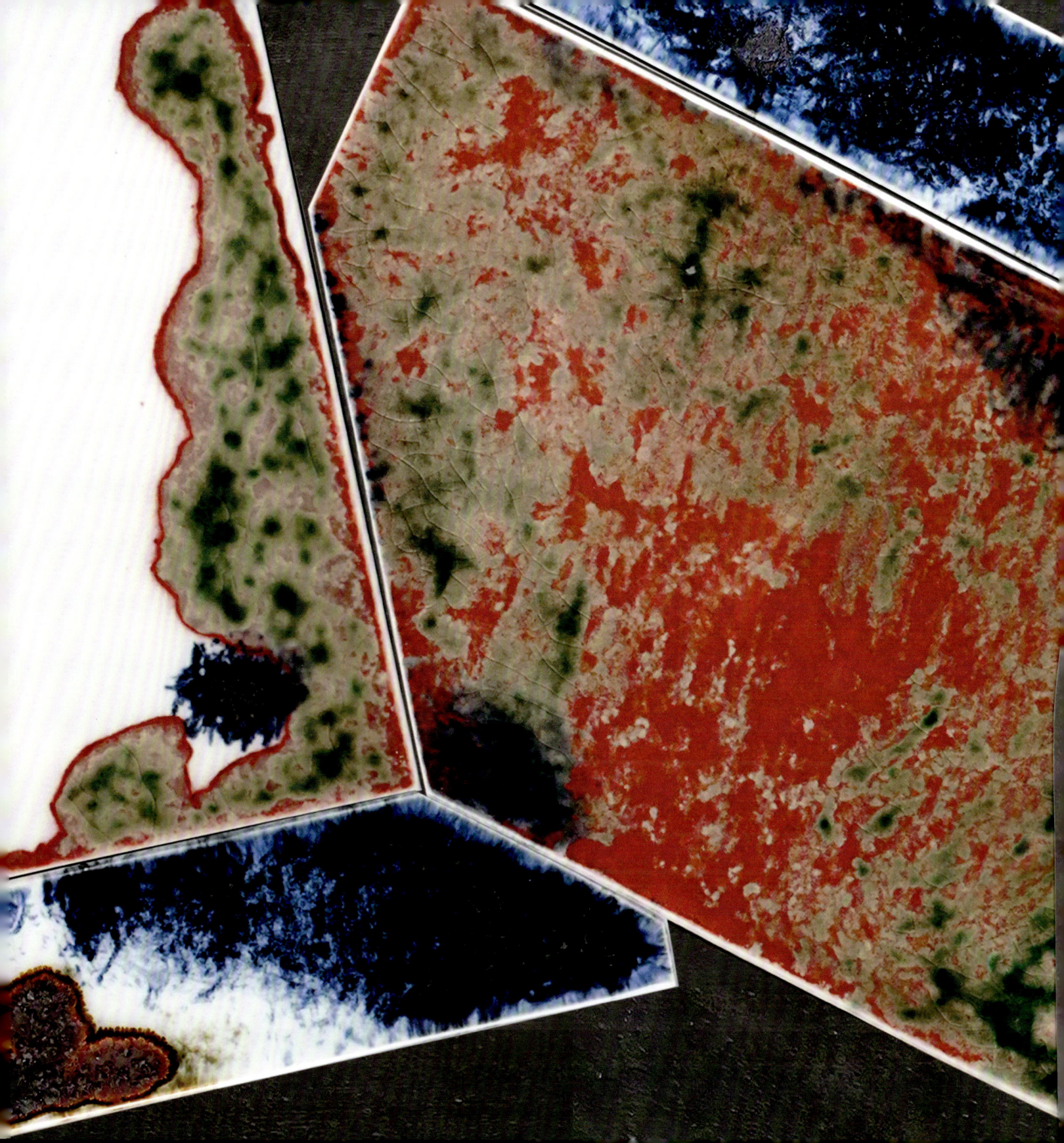

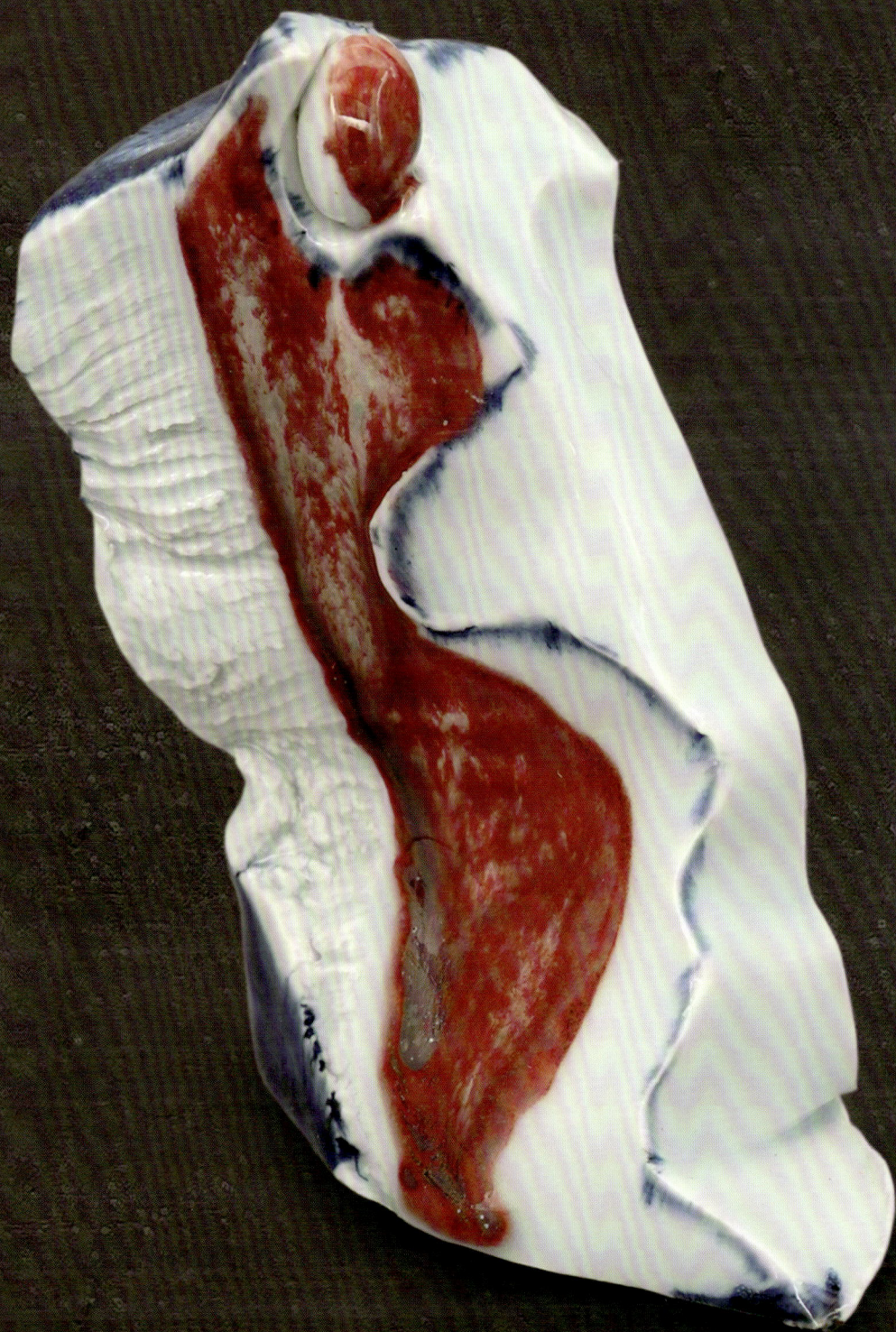

长：8cm，宽：12cm，高：17cm

洪劲松，感觉好、善品悟，施釉、装窑工艺高手。一丝不苟，精益求精，尤精于青花釉里红罩釉技艺，善于根据色料厚薄、画风工写确定釉层厚薄，精微只在毫厘之间。经他手出窑的釉里红多有"醉意"，业界称其为"神厚度"。

　　Hong Jinsong has a keen sense and is good at appreciation and comprehension. He is an expert in glazing and kiln loading techniques. he is meticulous and striving for perfection, and particularly proficient in the technique of applying the over glaze for blue and white with underglaze red. He is adept at determining the thickness of the glaze layer according to the thickness of the color material and whether the painting style is elaborate or free hand, with the precision lying within a hair's breadth. The underglaze red works fired from the kiln under his operation often have a charming "intoxicating" quality, and the industry refers to him as the "master of glaze thickness".

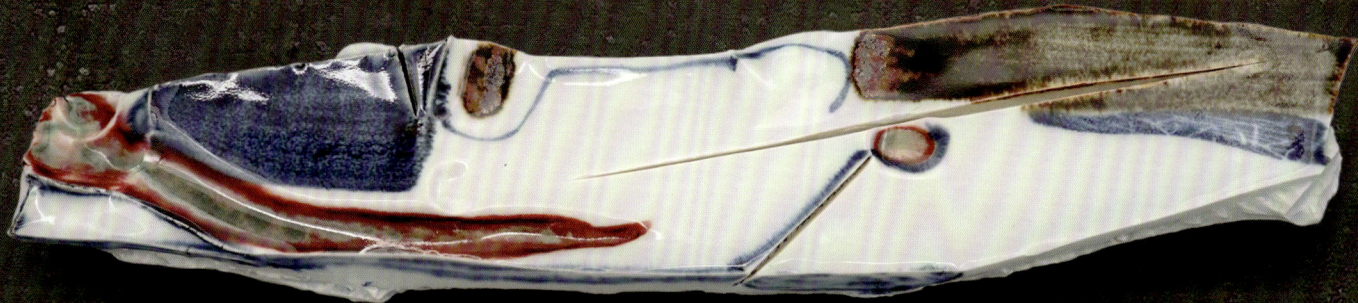

长：27cm，宽：10cm，高：6cm

长：27cm，宽：11cm，高：7cm

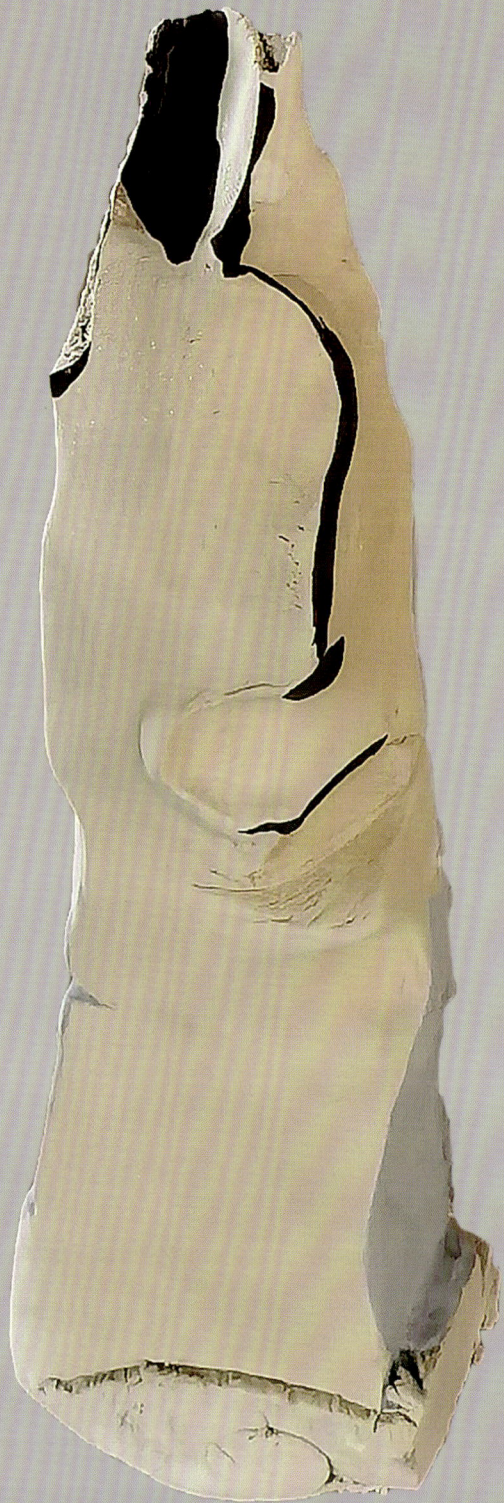

长：10cm，宽：10cm，高：25cm

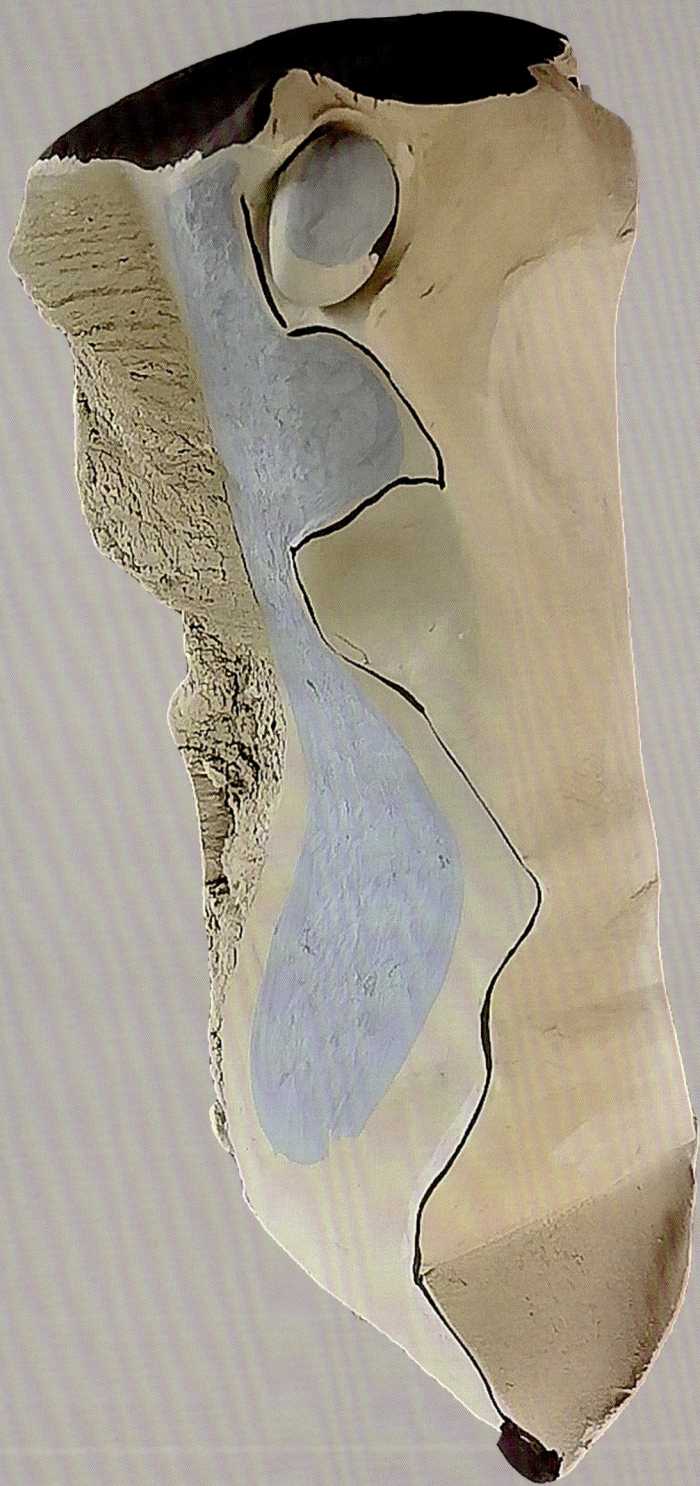

长：10cm，宽：10cm，高：21cm

长：10cm，宽：10cm，高：21cm

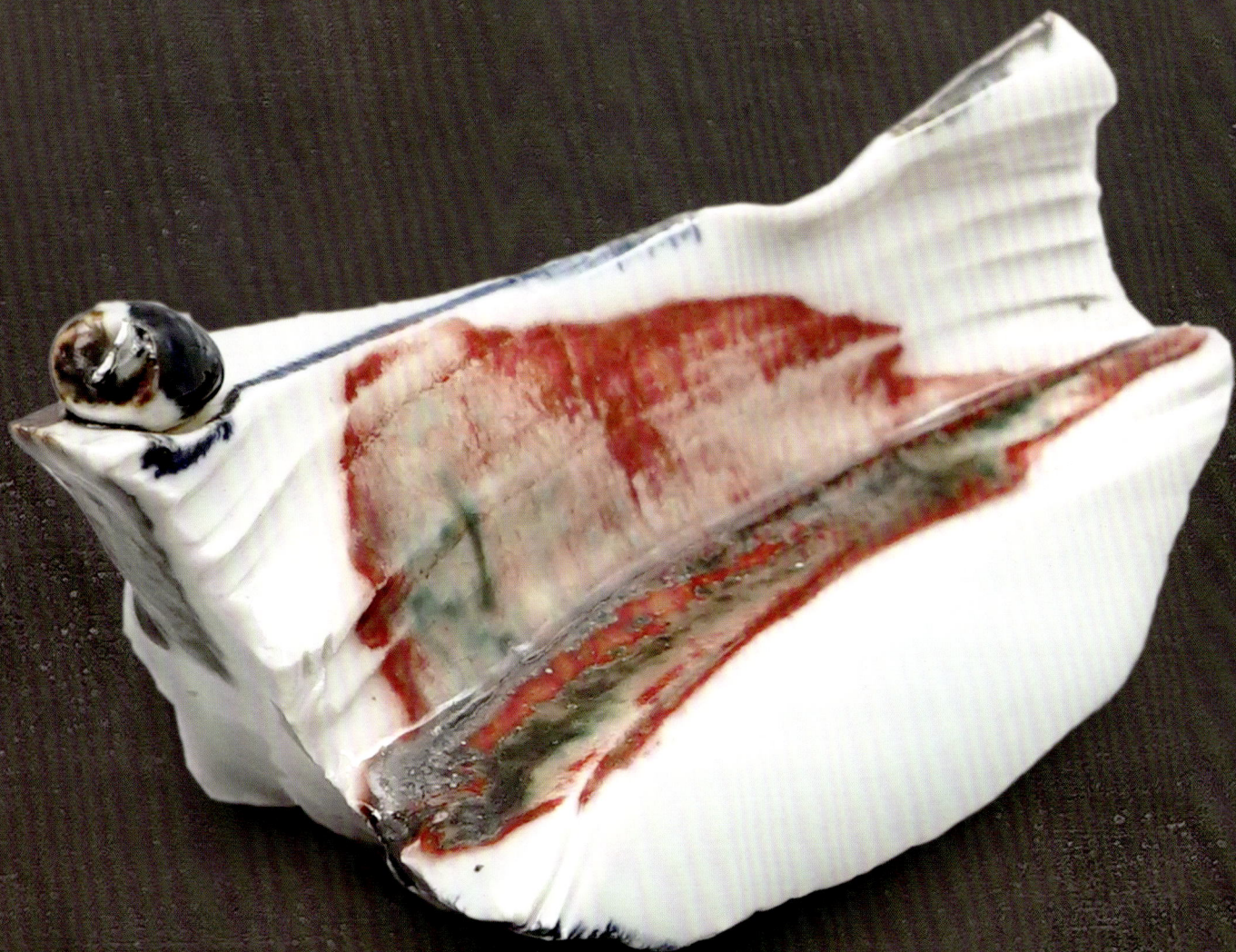

长：17cm，宽：8cm，高：11cm

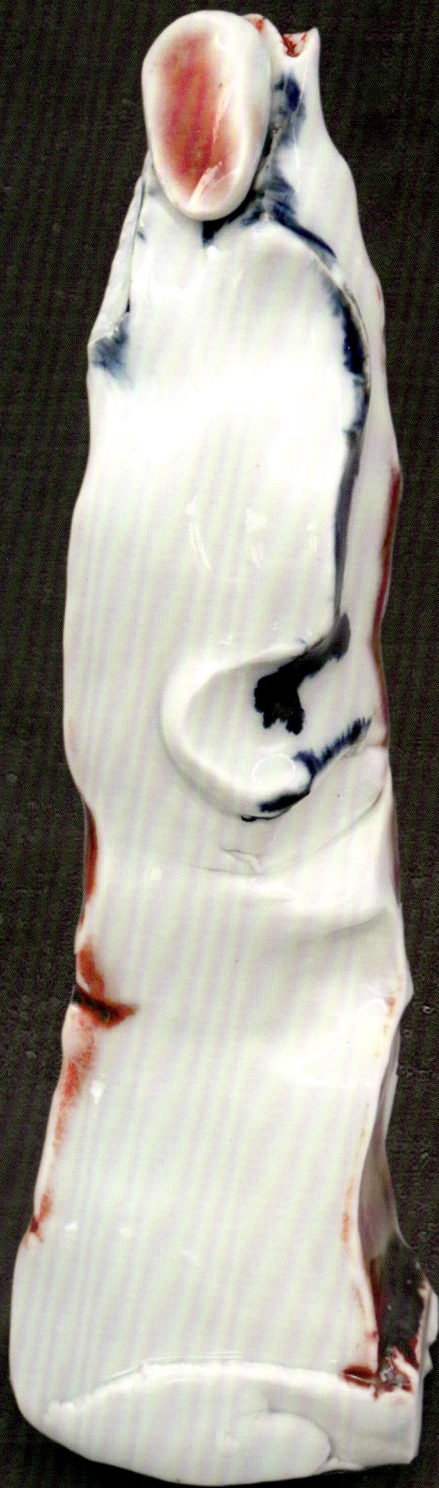

长：10cm，宽：8cm，高：21cm

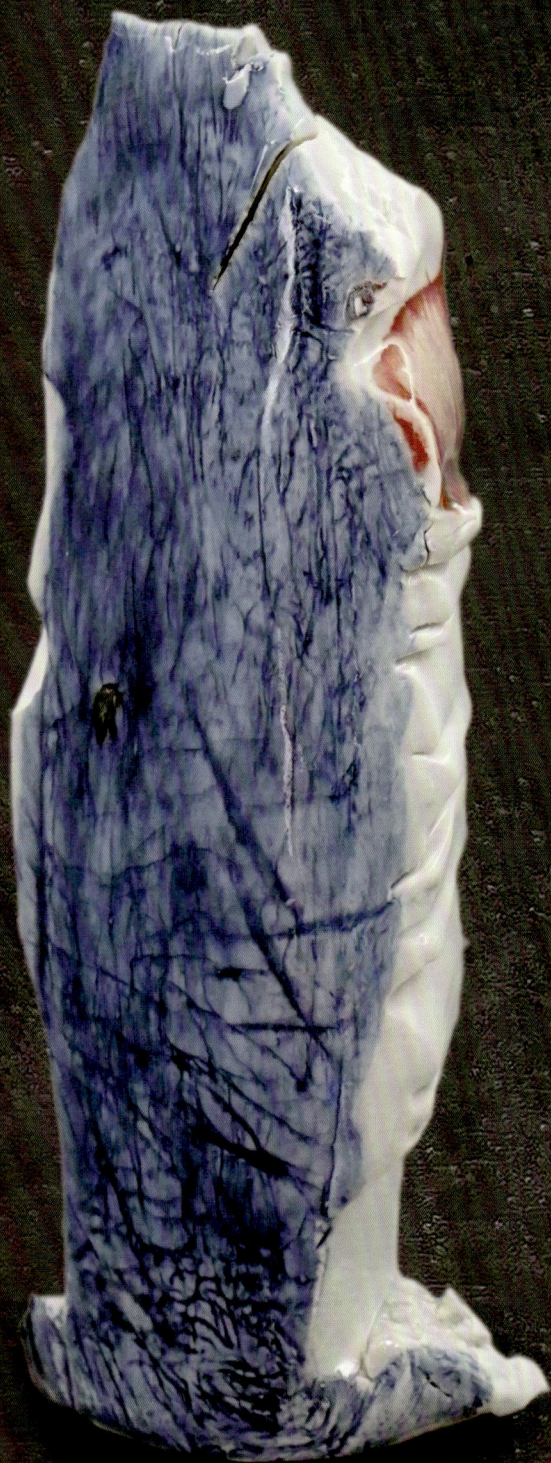

长：10cm，宽：9cm，高：21cm

长: 12cm, 宽: 10cm, 高: 21cm

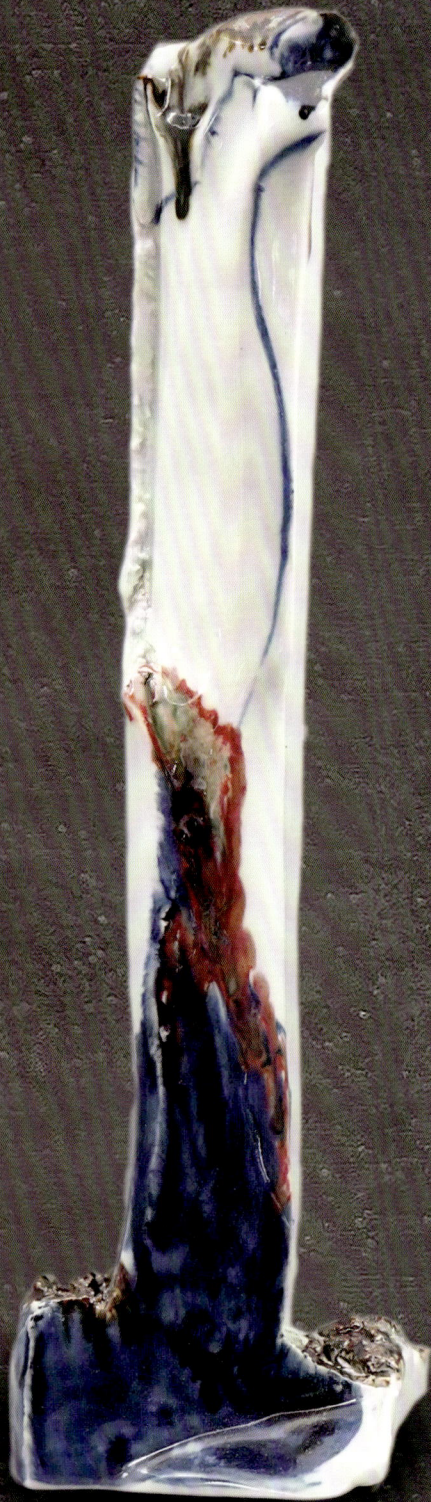

长：10cm，宽：10cm，高：26cm

长：180cm，宽：50cm

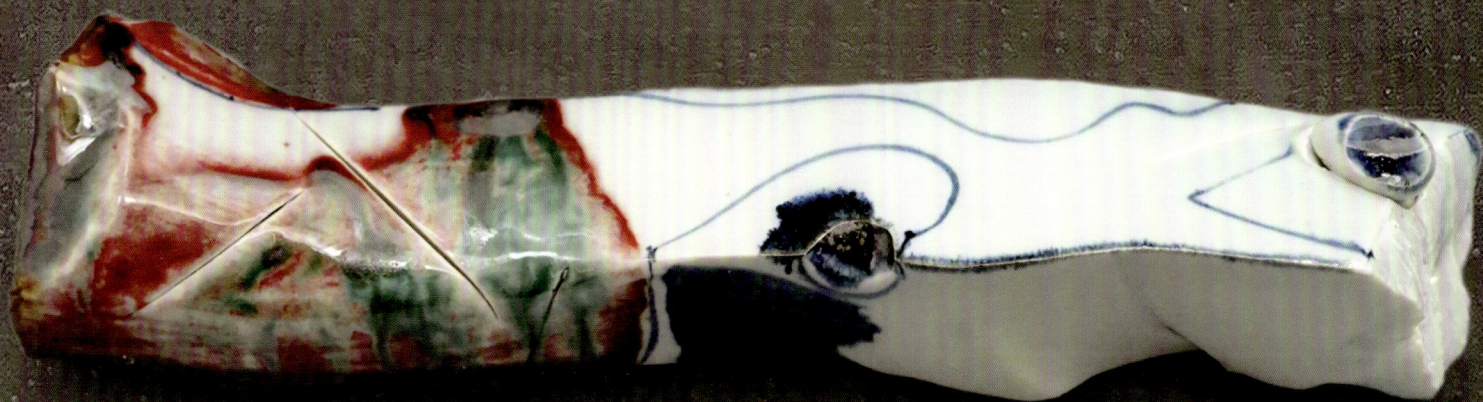

长：30cm，宽：10cm，高：8cm

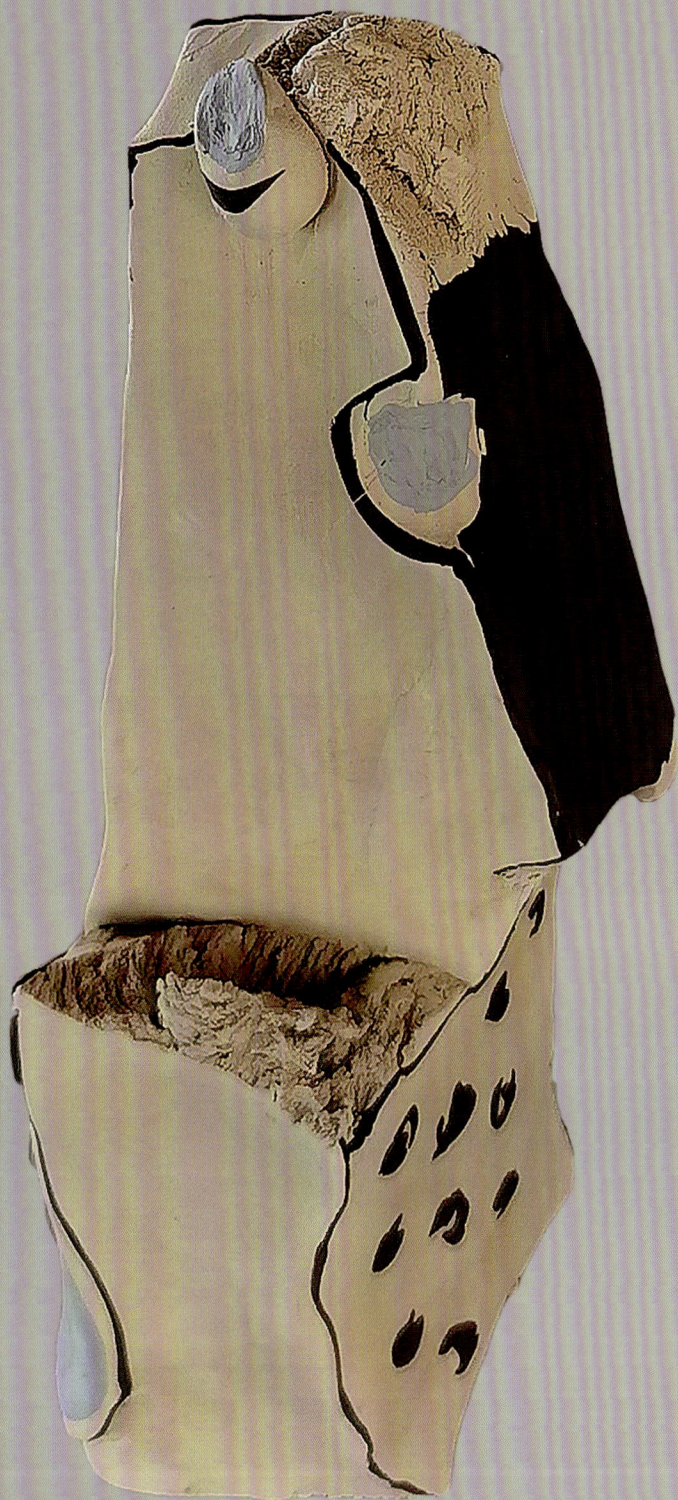

长：10cm，宽：10cm，高：21cm

长: 10cm，宽: 10cm，高: 21cm

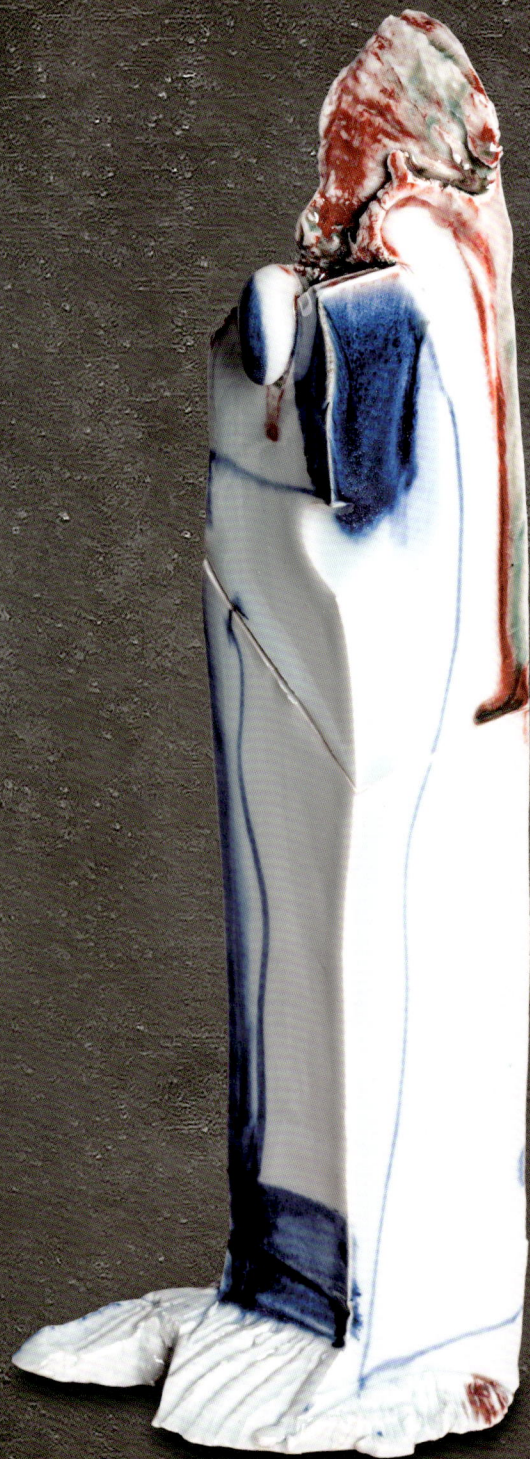

长：10cm，宽：12cm，高：27cm

长：10cm，宽：12cm，高：27cm

长：7cm，宽：7cm，高：32cm

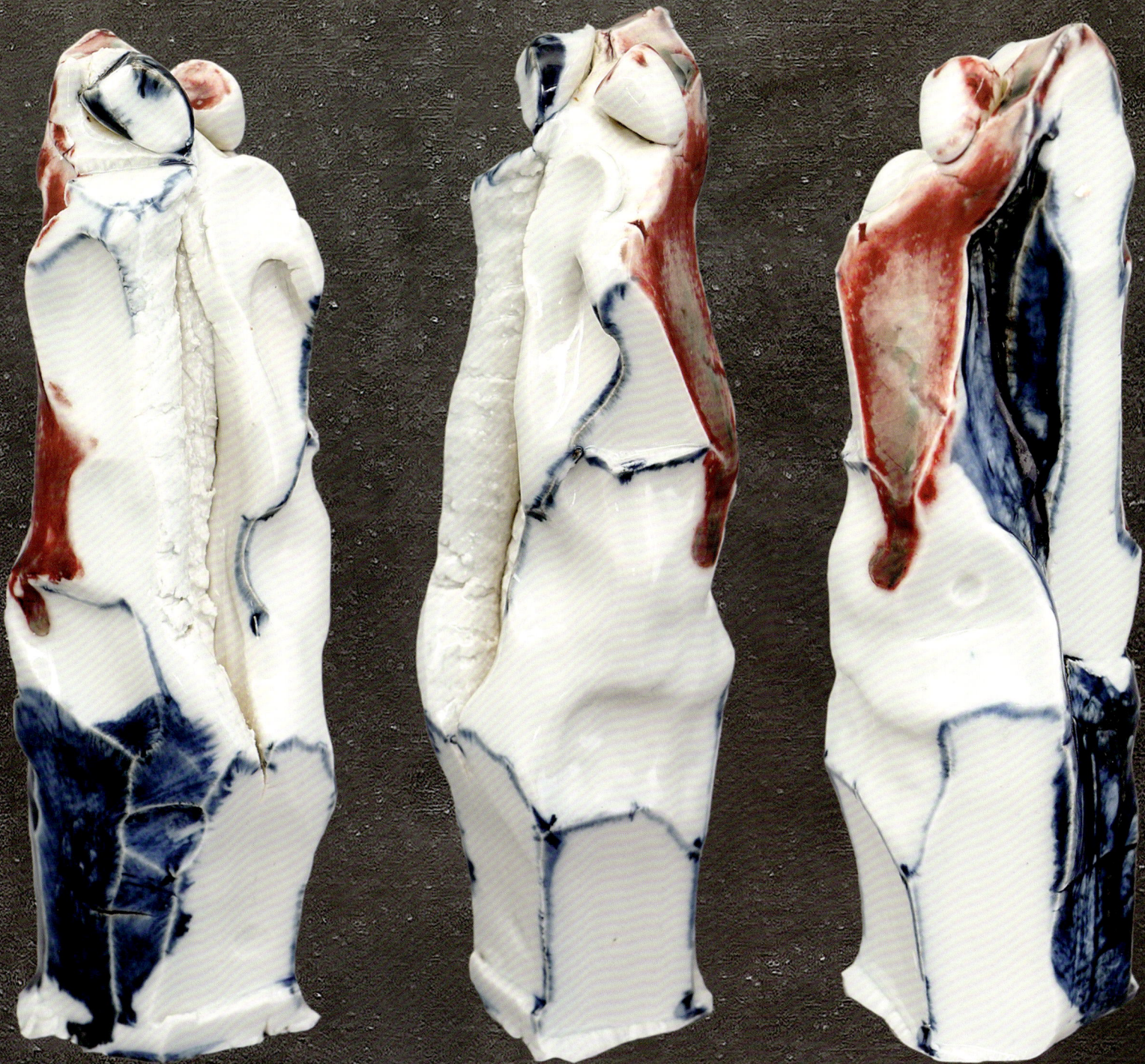

长：8cm，宽：8cm，高：23cm，数量：3

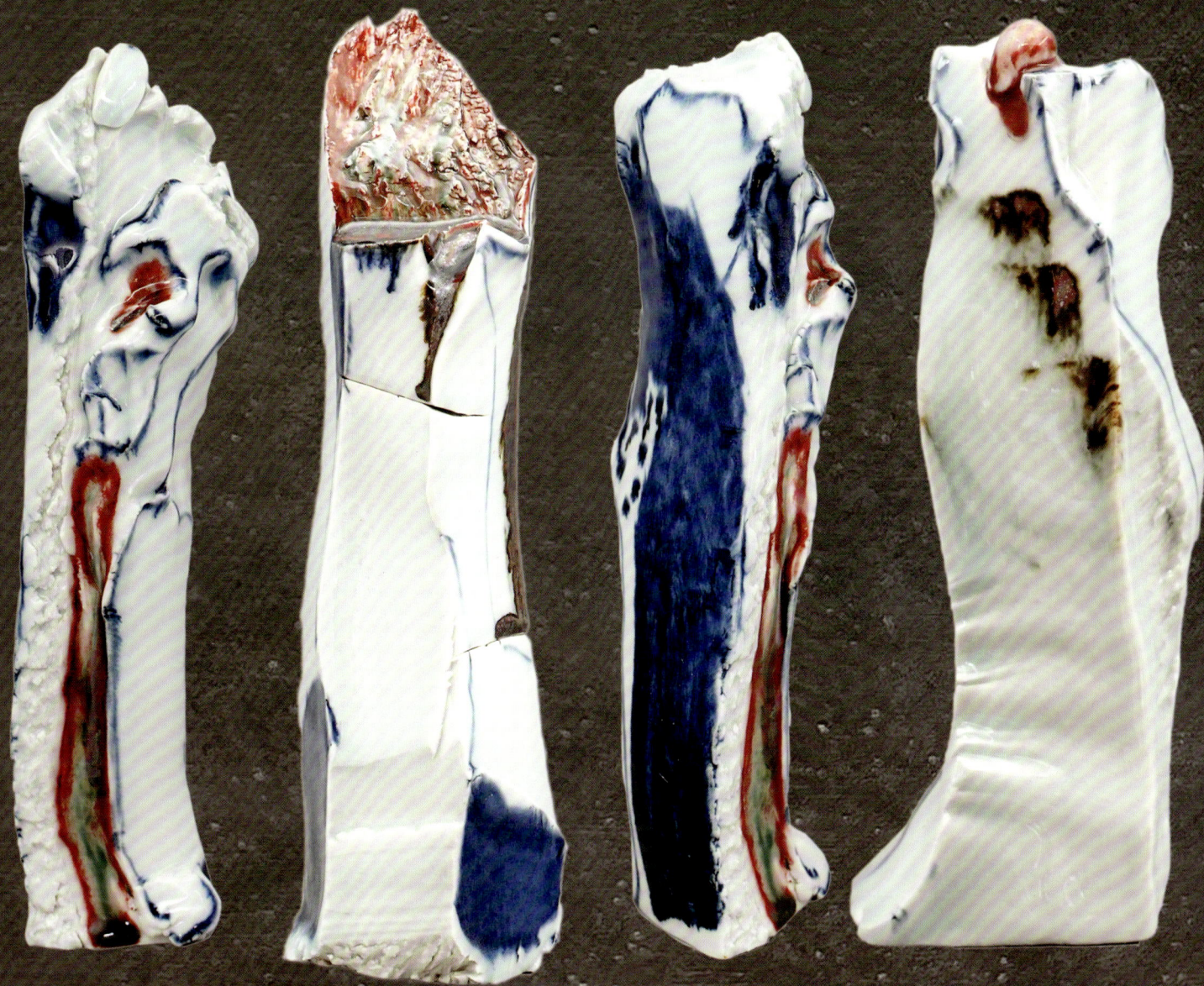

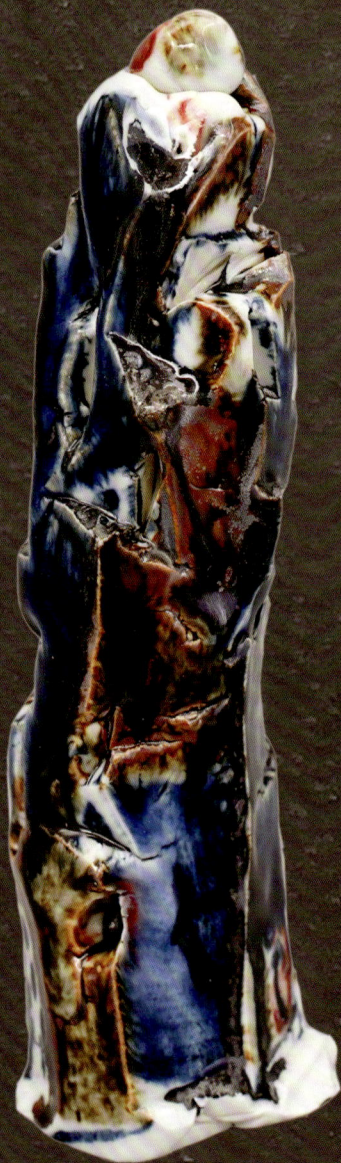

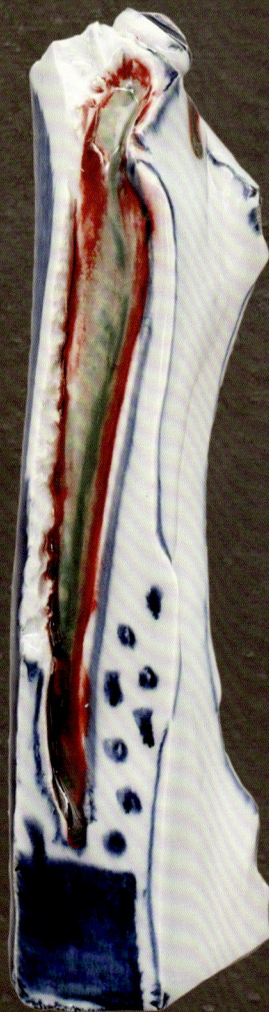

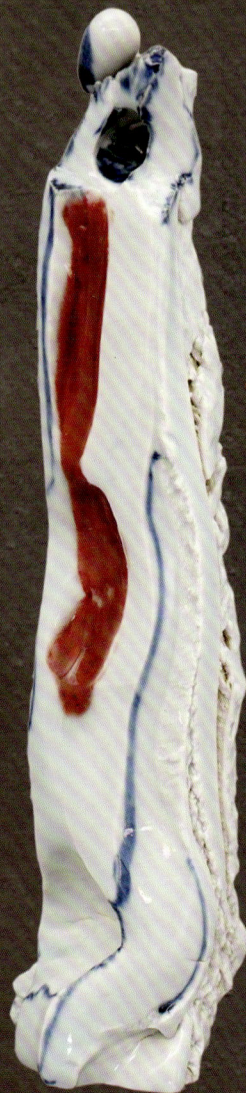

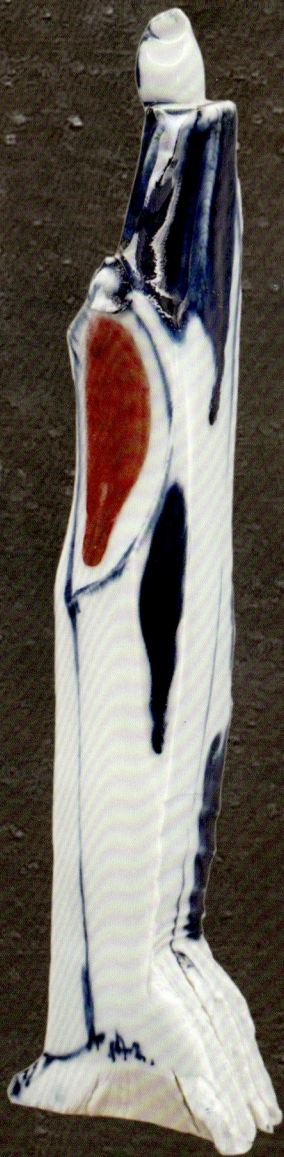

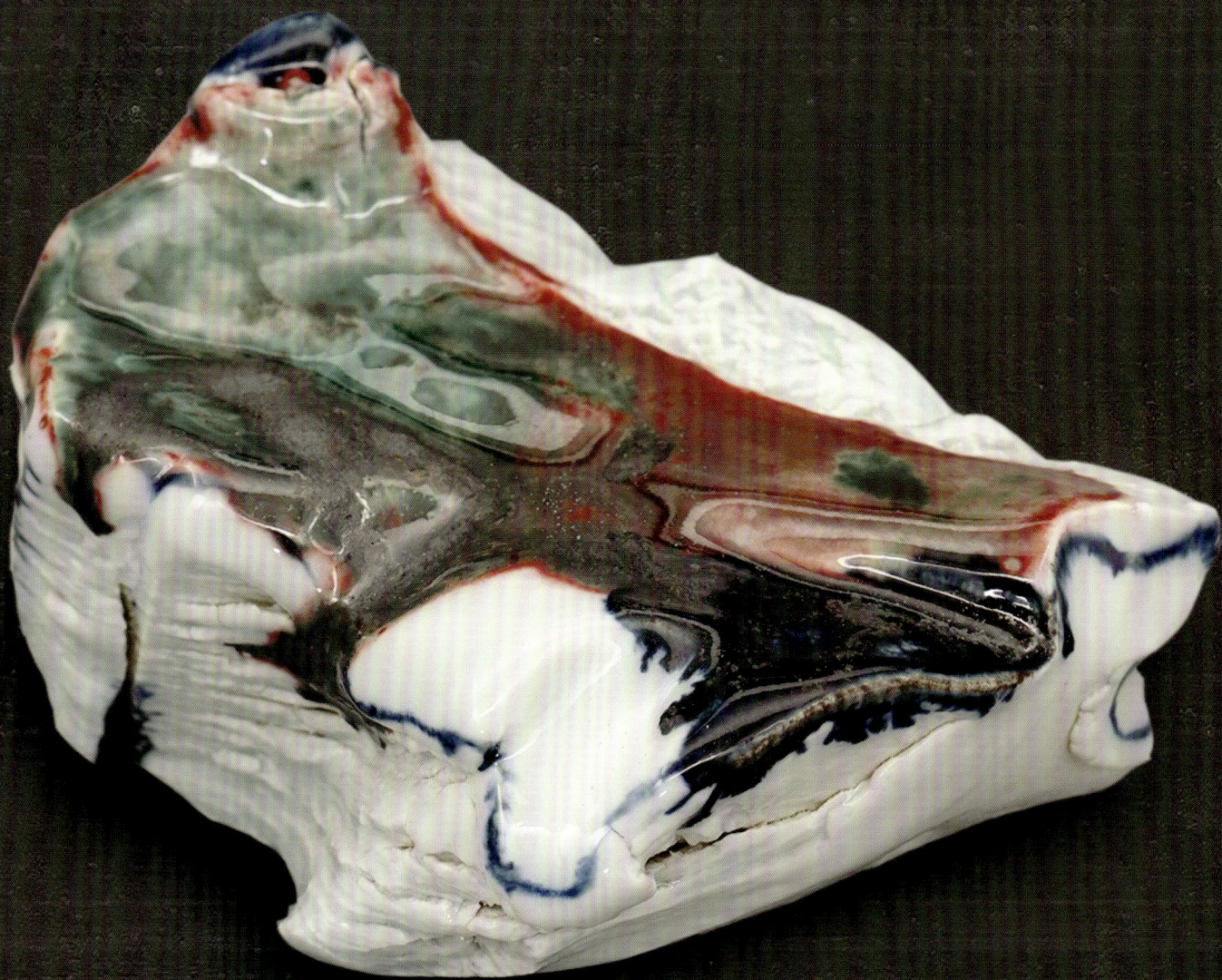

长：15cm，宽：15cm，高：10cm

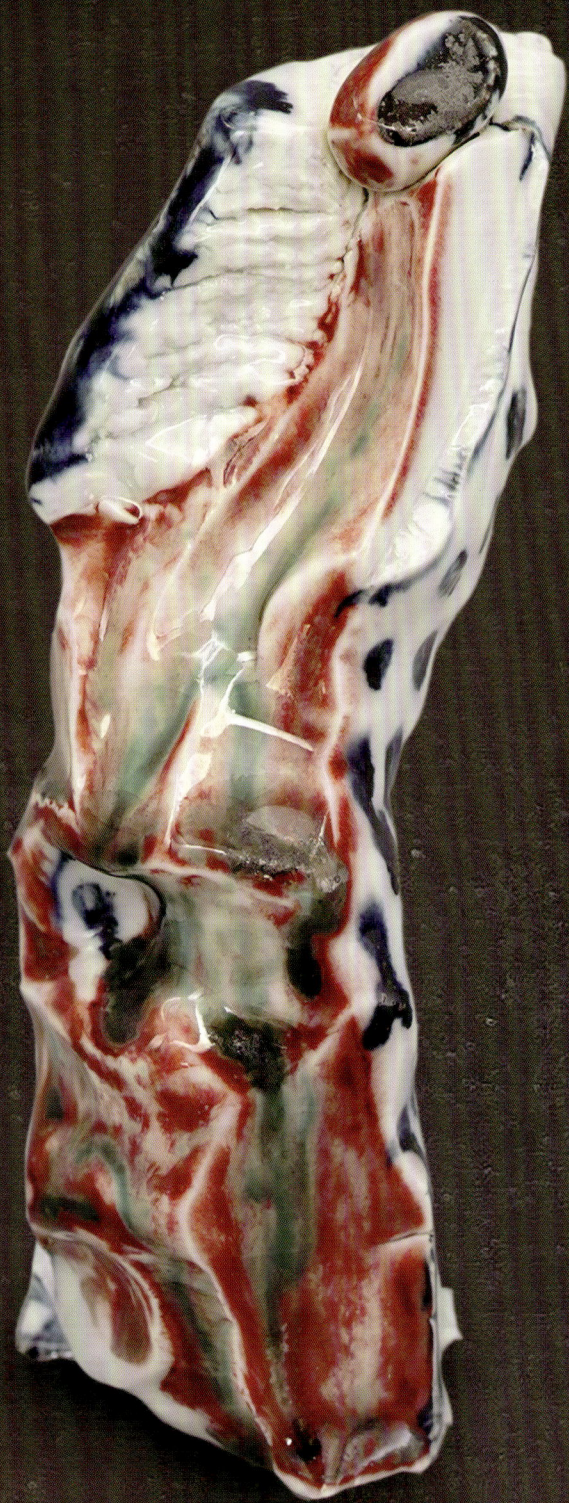

长：10cm，宽：10cm，高：25cm

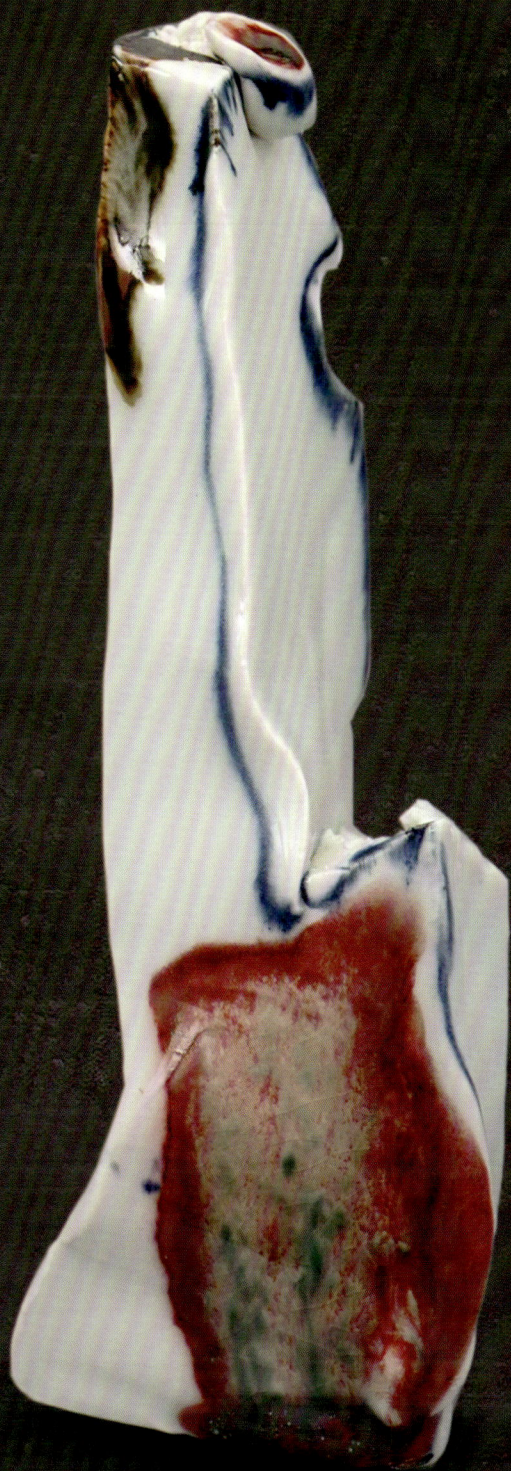

长：10cm，宽：10cm，高：26cm

肖颖　Xiao Ying

艺术家，正高级工艺美术师，中国美术家协会会员。

Artist, a senior craftsman at the highest level, and a member of Chinese Artists Association.

对于一个有信仰的人而言，我试图在画面中呈现出一种人与自然共生的状态，同时表达出对生命的敬畏和向往。在创作时，天气的阴与晴，空气的干与湿，气候的炎与寒，都能导致作品呈现不同。创作中，在做好所有的物理准备之后，我通过泼、洒、倒、撕等一系列动作，与所使用的媒材发生关系，确定自己已经融入空间、时间、温度、湿度等元素组成的场域，并留下痕迹或是存在的证明之后，才开始关照造型、色彩，关照画面本身，直至最后完成作品。

For a person with faith, I attempt to present a state of symbiosis between humanity and nature in my paintings, while expressing reverence for and yearning towards life. During the creative process, the cloudy or sunny weather, the dry or humid air, and the hot or cold climate can all lead to different presentations of the work. In my creation, after making all the physical preparations, I interact with the media through a series of actions such as splashing, sprinkling, pouring, and tearing. Only after confirming that I have integrated into the field composed of elements like space, time, temperature, and humidity, and left marks or proof of my existence, do I start to pay attention to the shape, color, and the painting itself, until finally completing the work.

长：120cm，宽：120cm

长：10cm，宽：12cm

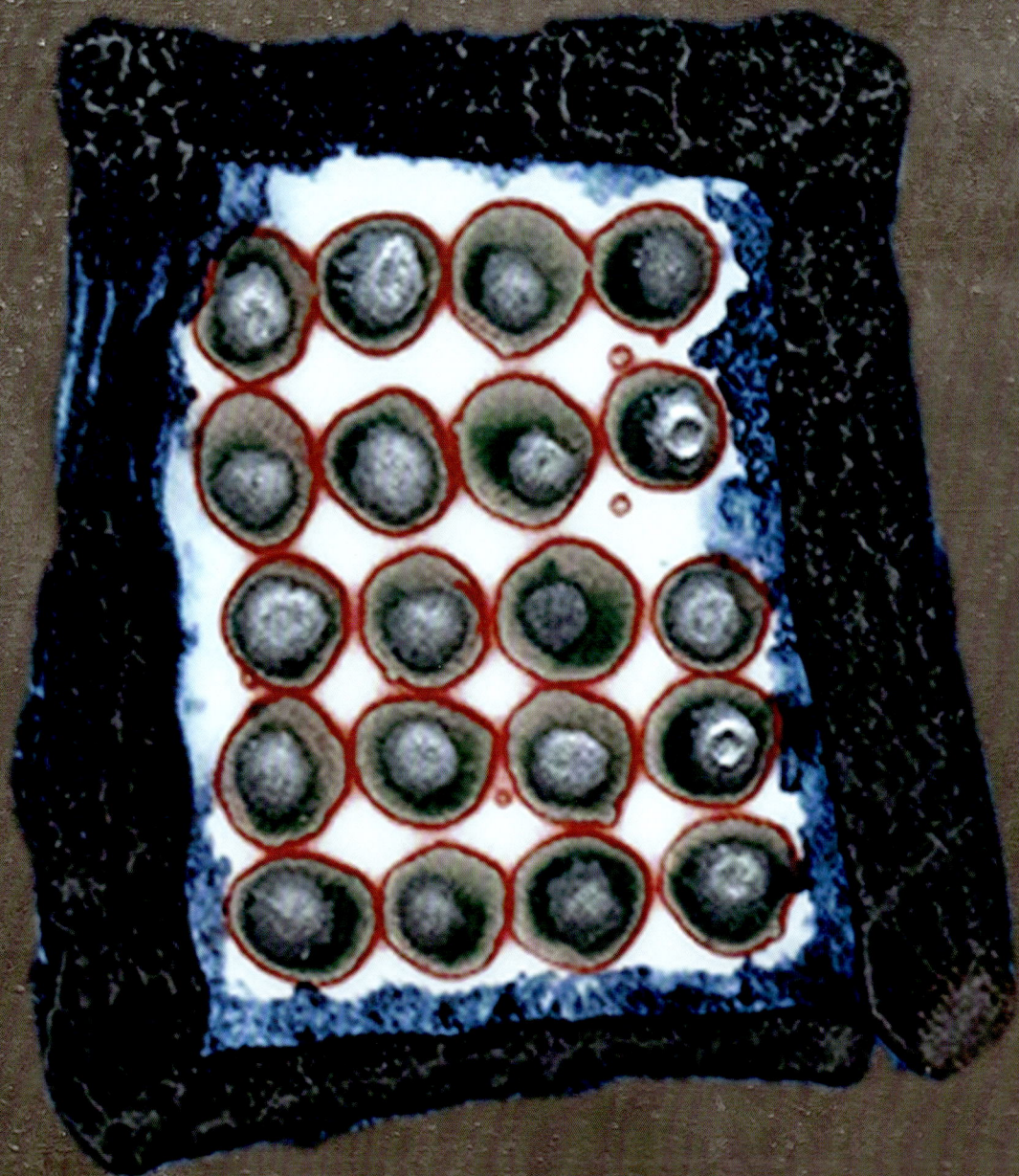

长：10cm，宽：12cm

长：10cm，宽：12cm

长：10cm，宽：12cm

长：10cm，宽：12cm

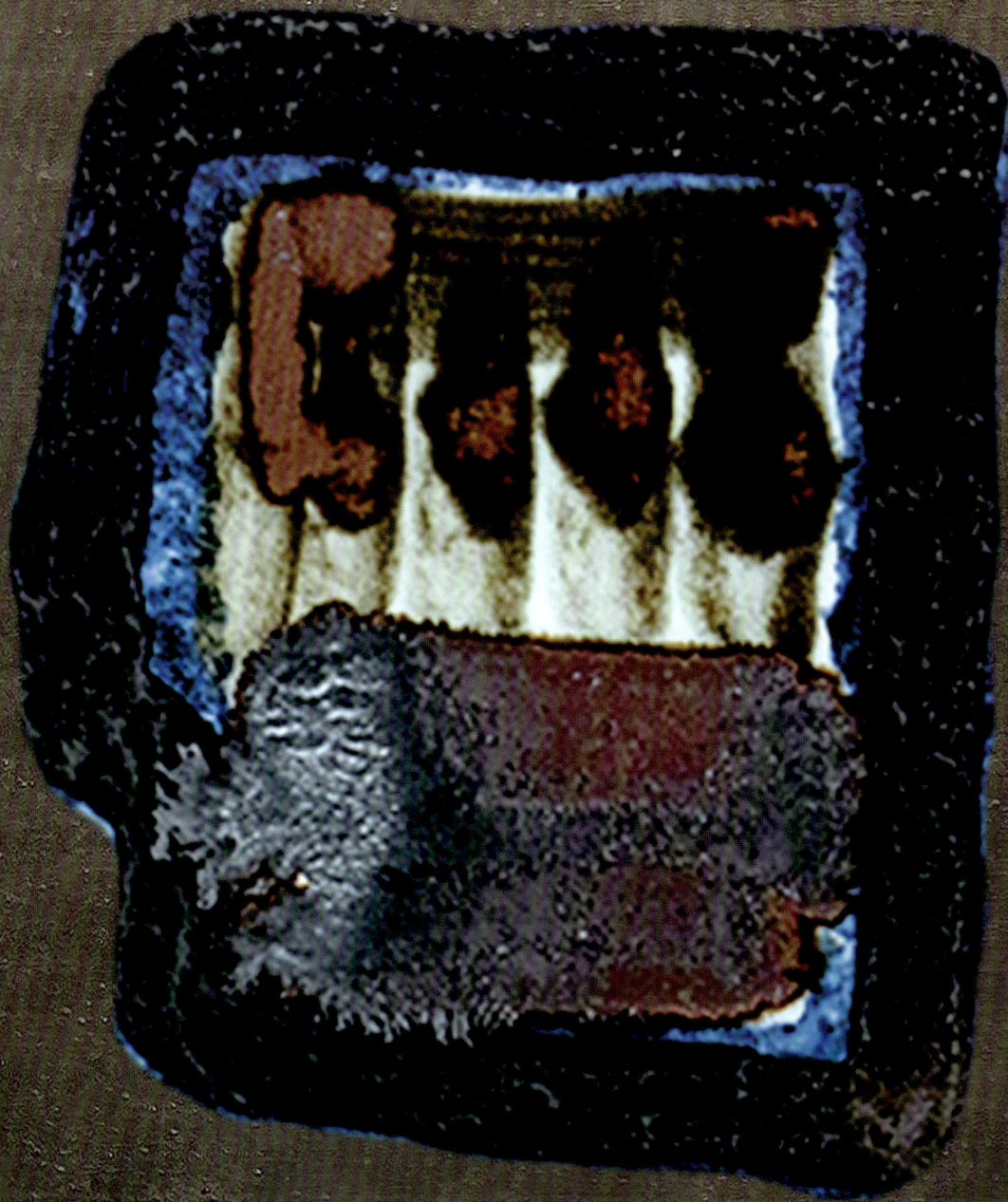

长：10cm，宽：12cm

直径：51cm，高：8cm

直径：51cm，高：8cm

171

长：120cm，宽：60cm

直径：51cm，高：8cm

长：120cm，宽：60cm

直径：51cm，高：8cm

175

长：50cm，宽：10cm，数量：8

长：50cm，宽：10cm，数量：8

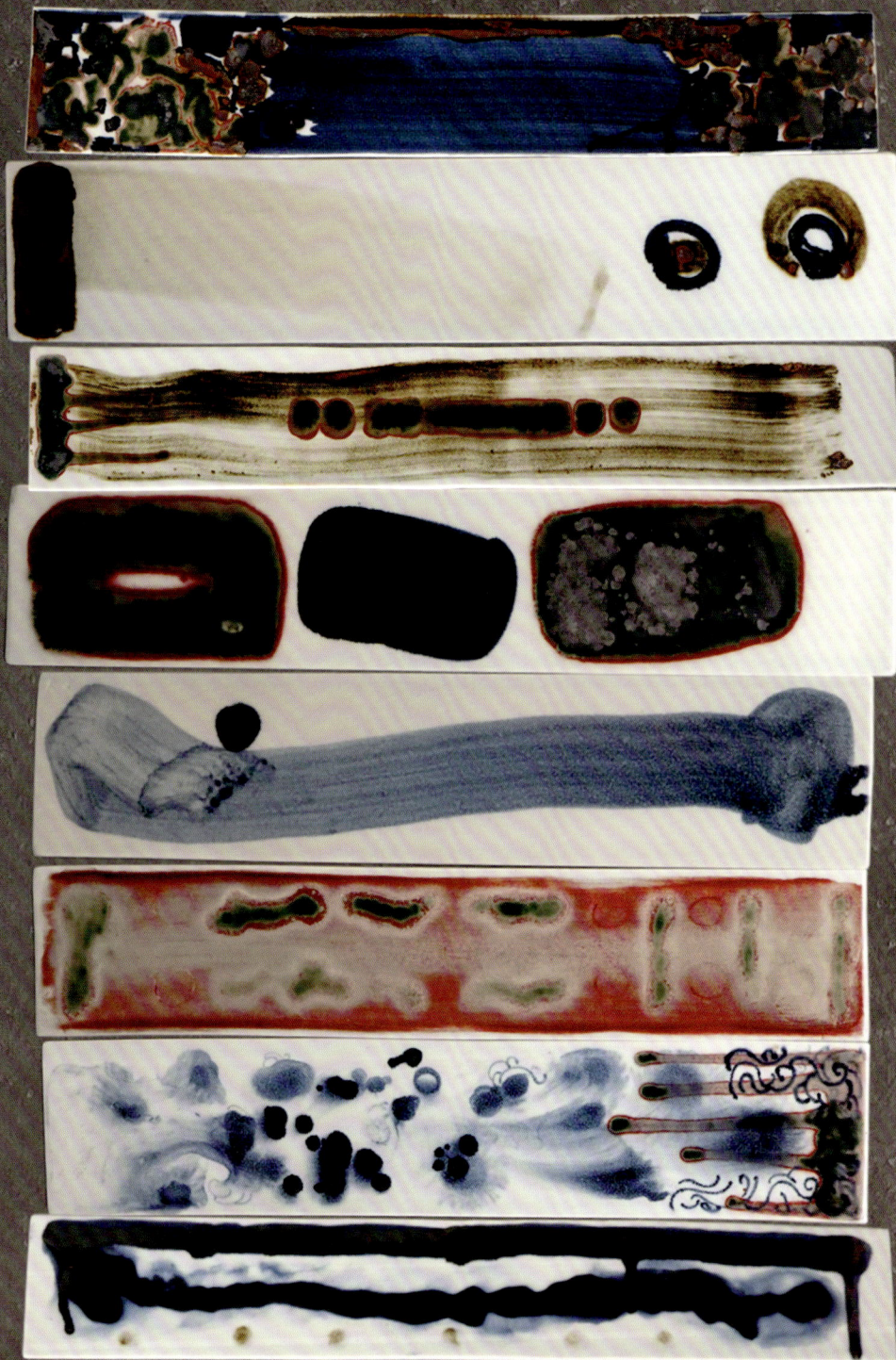

长：50cm，宽：10cm，数量：8

长：50cm，宽：10cm，数量：9

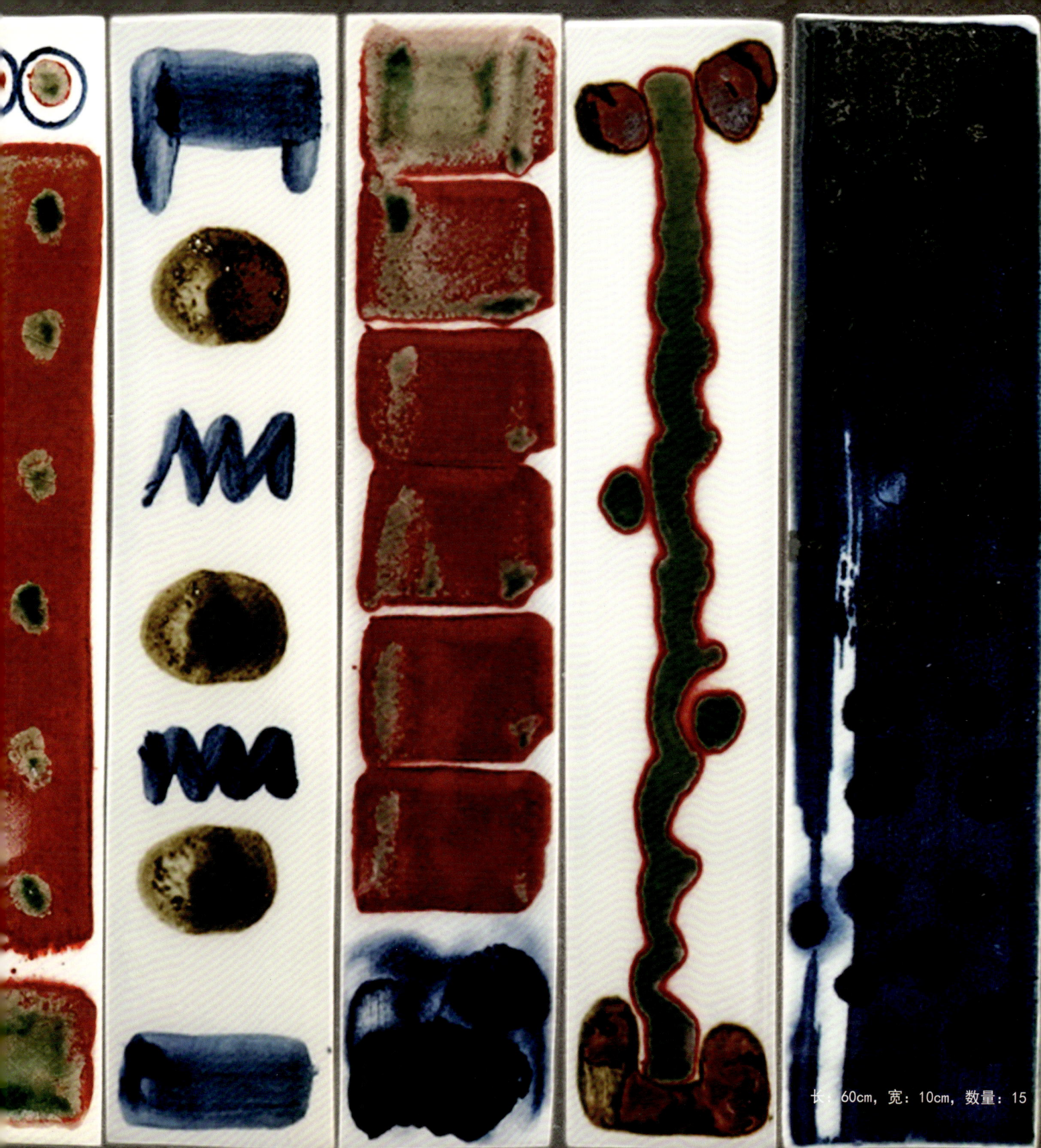

长：60cm，宽：10cm，数量：15

直径：51cm，高：8cm

黄利华，1969年10月出生于浙江金华，本科毕业于华北电力大学电力系，广西大学商学院工商管理硕士，曾获广西五一劳动奖章，六堡茶行业实力派领军人物。

Huang Lihua was born in Jinhua, Zhejiang in October 1969. He graduated from the Department of Electric Power at North China Electric Power University with a bachelor's degree and obtained a Master of Business Administration from the Business School of Guangxi University. He has won the Guangxi May 1st Labor Medal and is a leading figure in the Liubao tea industry with strong capabilities.

喝茶常常会有人讲所谓"茶道"。但从中国文化出发，我们只谈茶文化，不会随意地上升到"道"的层面，就像中国人写书，后世敢以"经"称的极少。其实中国人不会认同日本茶道是"茶道"，因为中国是茶叶的祖国，中国尚且无茶道，从中国传承的日本茶道当然也不能代表茶之道。但我们的茶生活和茶体验中常常有茶道的存在。这种看不见的规矩左右着我们，只喝纯茶，花茶、添加茶等不以茶的待遇待之。我们喝茶讲究层次和香气，要喝茶的真香。我们时时能感受到茶道的存在，这种存在，推动着中国茶叶和茶人的进步，也是中国茶一直走在世界前列、引领世界茶叶发展的指南针。

　　一个老生常谈的问题，说中国所有的茶企加在一起比不上一个立顿，以此说明中国的茶企很落后、规模小，中国企业不讲标准等等。如果比销量和产量确实如此，但立顿入茶道了吗？答案是否定的。单看中国众多口感和门类繁多的茶品，就是茶文化的骄傲。如果中国茶业只剩十家立顿，那一定是中国茶的悲剧，也是世界茶业的悲剧，那时中国一定不会再有人谈茶道了。

　　茶不是生活必需品，但丰富多样，有更高的文化价值。从禅的角度出发，大千世界没有两片相同的树叶，这就是中国的茶，中国的茶道。

　　中国茶如同中国的餐饮，不仅仅满足于口腹之欲。就像中国的绘画，一千年前已经有大写意了，而西方到印象派才开始写意。我们的茶已经领先了几千年，而我们却以最不入流的西方茶叶标准来衡量我们自己。

　　六堡之道也是六堡茶之道。六堡从来不具茶文化上的话语权，如同在西方规则下的中国茶一样。它产生在南方的崇山之中，没有文人墨客歌颂，没有皇帝高僧的垂青，但它依然不屈地生长，因为它契合了茶道。它不是在与世隔绝的山洞中得到一本秘籍的武学奇才，它只是一个日复一日练着《易筋经》的扫地小和尚，靠时间和坚持获得成就，也许它自己也不曾预料到会有一鸣惊人的时刻，这就是六堡茶。

　　中国的茶道，六堡茶当为杰出代表之一。

六堡茶如同走出深山的侠客，它获得自己不曾预料的好评，于是成为名茶。这可以是一部戏曲的剧本，但艺术往往来自现实。

六堡茶之所以成为名茶有三个原因：

一、时代选择了它。六堡茶成名的清代，民间疾苦加剧，雅致生活受冲击，精致的茶不再成为主流，民间需要六堡茶这种好喝、便宜和功效明显的茶，六堡茶应时代的要求走出了深山。当时在广州的茶楼都有六堡茶供应。得益于粤商和广州这个大码头，六堡茶开始沿河而下，跨洋出海。在1860年至1950年的近百年间，南洋开发的热潮兴起，南方沿海各省有近1500万人下南洋寻找机会。咸丰、同治年间，厦门、汕头、广州、澳门和香港都成了苦力的贸易中心。南洋华工们主要在矿山和橡胶园干活，工作环境迫使华工们只能喝从家乡带去的可以祛湿、调理肠胃的六堡茶。因此，六堡茶在南洋很多矿山也被称为"养命茶"。

二、价格便宜且没法仿制。六堡茶方便喝，可以像煮开水一样煮开喝，也可以焖着喝，十分方便。并且喝了以后很舒服，比大部分茶适口性更好，刺激性不强，人人都可以喝。

三、有很好的药用功效。六堡茶除瘴祛痢，可以治痢疾，缓解消化不良。劳动阶层，尤其在异国他乡的人，很容易因水土不服引起身体不适，喝六堡茶就可以解决，这是缺医少药时代的最佳保健品。

以上三点只是想说明六堡茶在清中叶后的成名，是时代的必然。人们都是在事后寻找成功的原因，内在逻辑说不清也道不明，这也是所谓"六堡之道"吧。

黄利华

2025.2.7

When it comes to drinking tea, some people often mention the so called "tea ceremony". However, from the perspective of Chinese culture, we only talk about tea culture and don't casually elevate it to the level of "Tao". Just as in China, few books written by later generations dare to be called "classics". In fact, the Chinese don't recognize the Japanese tea ceremony as the real "tea ceremony" because China is the birthplace of tea. Since there is no such thing as a "tea ceremony" in China, the Japanese tea ceremony, which is passed down from China, surely can't represent the true way of tea either. But the tea ceremony often exists in our tea drinking lifestyle and experience. These invisible rules govern us. We only drink pure tea and don't treat scented tea, added ingredient tea, etc. as real tea. We pay attention to the layers and aroma of tea, seeking to taste its true fragrance. We can constantly feel the existence of the tea ceremony, which promotes the progress of Chinese tea and tea related people. It also serves as a compass that keeps Chinese tea at the forefront of the world and leads the development of the global tea industry.

There is a common issue: it's said that all Chinese tea enterprises combined can't match Lipton, suggesting that Chinese tea enterprises are backward, small scale, and don't adhere to standards. In terms of sales volume and production, this may be true. But has Lipton grasped the essence of the tea ceremony? The answer is no. Just looking at the numerous Chinese teas with diverse tastes and varieties, they are the pride of tea culture. If there were only ten Liptons left in the Chinese tea industry, it would be a tragedy for both Chinese and global tea. At that time, no one in China would talk about the tea ceremony anymore.

Tea isn't a necessity of life, but it's rich and diverse, with higher cultural value. From a Zen perspective, no two leaves in the world are the same. This is Chinese tea and the Chinese tea ceremony.

Chinese tea is like Chinese cuisine; it's not just about satisfying hunger. Just as Chinese painting had freehand brushwork a thousand years ago, while the West only started with impressionist freehand painting. Our tea has been leading for thousands of years, yet we measure ourselves by the most unrefined Western tea standards.

The way of Liubao is also the way of Liubao tea. Liubao tea has never had a say in tea culture, just like Chinese tea under Western rules. It originated in the mountains of the south. Without praise from literati or favor from emperors and eminent monks, it still grows unyieldingly because it conforms to the tea ceremony. It's not a martial arts prodigy who gets a secret book in a secluded cave. Instead, it's like a little novice monk who sweeps the floor and practices the "Yi Jin Jing" day after day, achieving success through time and perseverance. Maybe it never expected to become famous overnight. This is Liubao tea.

Among the Chinese tea ceremony, Liubao tea is one of the outstanding representatives.

Liubao tea is like a knight errant emerging from the mountains. It has received unexpected acclaim and thus become a famous tea. This could be the plot of an opera, but art often comes from reality.

There are three reasons why Liubao tea has become a famous tea:

First, the times have chosen it. During the Qing Dynasty when Liubao tea became well known, the hardships of the common people increased, and elegant life was impacted. Exquisite tea was no longer the mainstream. The common people needed Liubao tea, which was delicious, inexpensive, and had obvious effects. Responding to the needs of the times, Liubao tea emerged from the mountains. At that time, Liubao tea was available in teahouses in Guangzhou. Thanks to Cantonese merchants and the port of Guangzhou, Liubao tea started to go downstream along the river and across the ocean. From 1860 to 1950, during the nearly hundred year upsurge of the development of Southeast Asia, nearly 15 million people from the southern coastal provinces went to Southeast Asia to seek opportunities.

During the reigns of Emperor Xianfeng and Emperor Tongzhi, Xiamen, Shantou, Guangzhou, Macau, and Hong Kong became trading centers for coolies. The Chinese laborers in Southeast Asia mainly worked in mines and rubber plantations. The working environment forced them to drink Liubao tea brought from their hometown, which could dispel dampness and regulate the stomach. So, Liubao tea was also called the "life sustaining tea" in many mines in Southeast Asia.

Second, it's inexpensive and hard to imitate. Liubao tea is convenient to drink. It can be boiled like water or steeped, which is very convenient. After drinking it, people feel comfortable. It has better palatability than most teas, with less irritation, so everyone can drink it.

Third, it has good medicinal effects. Liubao tea can dispel miasma and relieve dysentery, treating dysentery and alleviating indigestion. For the working class, especially those in foreign lands, it's easy to get sick due to unacclimatization, and drinking Liubao tea can solve this problem. It was the best health care product in the era when medical resources were scarce.

The above three points are merely intended to illustrate that the rise to fame of Liubao tea after the mid Qing Dynasty was an inevitable outcome of the times. People always look for the reasons for success after the fact, and the internal logic is hard to clearly explain. Perhaps this is the so called "way of Liubao".

Huang LiHua
February 7th, 2025